DIVIDED WE FALL

The Confederacy's Collapse

From Within

A State-by-State Account

Calvin Goddard Zon

Printed in the United States of America

First Printing, 2014

ISBN: 978-0-692-29439-0

ZonBlaydes Publishing

ZonBlaydes@aol.com

Cover illustration by Roy Comiskey

ACKNOWLEDGEMENTS

Special thanks and gratitude to my wife Laurel, without whose technical expertise and encouragement this project would have been impossible.

Special thanks to William Connery for his eagle-eyed proofreading.

Special thanks to Richard Griffin for his suggestions regarding the manuscript.

Special thanks to Jack Berkley for his suggestions regarding the manuscript's introduction.

Special thanks to Ken Childs for sparking interest among Lincoln Forum authors.

TABLE OF CONTENTS

By state in order of secession

INTRODUCTION

Following the disastrous Union defeat at Fredericksburg in December 1862, 2nd Lt. Henry Perkins Goddard of the 14th Connecticut Infantry, my great-grandfather, wrote to his older brother, "I have little hope that the South will ever be conquered. The reason is that they are united, and are fighting on their own soil. We are divided: half the army hates the Negro, the other half believes in abolition. Too many in the North are wearying of the war." (*The Good Fight That Didn't End: Henry P. Goddard's Accounts of Civil War and Peace*, Calvin Goddard Zon, Editor, University of South Carolina Press, 2008).

Goddard was right about division and war weariness among Northerners, but he was dead wrong about a united South. The story of antiwar Northern Copperheads, the New York draft riot, and peace candidate George McClellan's challenge to President Lincoln is well known. Contrary to popular lore, active and organized opposition to the war was considerably stronger in the South than in the North.

Opposition to secession was especially strong in the mountainous, mostly non-slaveholding sections of North Carolina, South Carolina, Tennessee, Georgia, Alabama, and Mississippi. Among the six states of the Lower South that were first to secede, only Texas and Louisiana put the question to voter referendum.

In the other four states, voters selected secession convention delegates mostly hand-picked by the slaveholding planter class, and to assure the outcome, threats and violence against Unionists were widespread. Voters in Arkansas and in the Upper South states of Virginia, North Carolina, and Tennessee opposed secession prior to the Battle of Fort Sumter and Lincoln's call for troops to put down the rebellion.

Early in the war, Southern Unionists formed secret peace societies, spy networks, and guerilla bands that resisted Confederate conscription, aided deserters and fugitive slaves, scouted for invading federal troops, and battled Confederate militia. Unionist bands of white Southerners controlled large areas of the Southern countryside by 1864, and

Unionists in several states defeated pro-Confederates in local elections.

As the war began, only a small minority of Southern anti-secessionists were outright abolitionists; the same might be said of opinion in the North, where early abolitionists were disdained as radicals.

Lacking sufficient voluntary enlistments, the Confederate Congress in April 1862, a year before the federal draft, enacted a draft law which exempted men of wealth who hired a substitute or paid a fee, and also those who owned 20 or more slaves. This caused much resentment among the three-fourths of Southerners who didn't own slaves and gave rise to the complaint, "Rich man's war, poor man's fight." (The same complaint was common in the North, where the 1863 draft law exempted those able to pay $300 or to hire a substitute.)

This class resentment was compounded by planters who devoted most of their land to the lucrative cash crops of cotton and tobacco while soldiers and their families grew increasingly hungry. In 1863, food riots in which gangs of women ransacked stores and warehouses occurred in several Southern cities, including the Confederate capital of

Richmond. Hunger at home and in the army contributed to a desertion rate that hampered the war effort.

General Robert E. Lee said in the fall of 1862 that desertion and straggling were important reasons for his army's failure to win victory at Antietam. In 1864, Confederate President Jefferson Davis admitted that one third of Confederate

soldiers were absent, most of them without leave. Defeat at Gettysburg and Vicksburg in early July 1863 turned the stream of desertion into a torrent.

Plagued by desertion and draft evasion, the Confederacy had only one quarter as many soldiers in the field by the end of 1864 as did the Union.

"Not only did the four million slaves identify with the Union cause, but large numbers of white Southerners came to believe that they had more to lose from a continuation of the war than from a Northern victory," wrote historian Eric Foner in *Civil War and Reconstruction*.

"Indeed, scholars today consider the erosion of the will to fight as important a cause of Confederate defeat as the South's inferiority in manpower and industrial resources. Even as it waged a desperate struggle for independence, the Confederacy was increasingly divided against itself," said Foner.

States rights, the basis on which the Confederacy was created in order to defend the institution of slavery, ironically became a major cause of its defeat. The Davis administration was sharply criticized by Southern state officials for trampling on states' rights with the conscription law, suspension of the writ of habeas corpus, and impressment of crops and livestock.

Some governors held back troops from the war effort, keeping them at home in defense of their own turf. A principal states' rights critic of President Davis was actually the vice president of the Confederacy, Alexander Stephens of Georgia, who spent more than two years of the war back in his home state, absent from Richmond.

The book includes state-by-state accounts of the activities and organizations of Southern Unionists and their leaders, who ranged from high-society spies and a turncoat Confederate general to dirt-farmer guerillas, propagandists, and politicians. Anti-Confederate activity was scattered across a surprising number of towns, cities, and counties in the 11 states of the Confederacy.

In Texas, Tennessee, North Carolina, Alabama, and Georgia—official monuments, plaques, and grave markers have been erected to honor martyred Southern Unionists and Southern soldiers who served in the Union army.

Some 100,000 white Southerners enlisted in the Union army and a state-by-state breakdown is included. South Carolina was the only state from which there were insufficient enlistments to form a Union battalion. It often took special courage to join federal ranks because capture could result in execution as a traitor and treatment of one's family by hometown neighbors could be harsh.

Of the 180,000 blacks who wore Union blue, about 150,000 were fugitive Southern slaves who made their way to Union lines.

A small preview sampling, in chronological order of secession:

In Charleston, **South Carolina**, the cradle of secession, the secret Union Association helped federal soldiers escape prison and return to Union lines.

In Jones County, **Mississippi,** dirt farmer Newton Knight, angry because wealthy slave owners could avoid the draft and because Confederate cavalry had seized his horse under the impressments law, deserted from the rebel army and went home to lead a guerilla band which fought off Confederate forces to the point that the county became known as the Free State of Jones.

In **Florida**, William Strickland, denied a furlough to care for his sick wife, deserted his Confederate cavalry regiment to lead a guerilla unit aided by local slaves who hid deserters and draft dodgers, funneled them food and supplies from the plantations, acted as spies, and often joined them in their raids.

The First **Alabama** U.S. Cavalry, recruited mostly from poor farmers in the mountainous northern part of the state, served as general Sherman's bodyguard, and the regiment seized every opportunity to sack plantations during Sherman's March to the Sea.

In Atlanta, **Georgia**, the Union Circle helped men escape the South to evade military service, acted as spies for invading federal forces, and provided food and medical treatment for prisoners of war.

In **Louisiana**, when Maj. Gen. Benjamin Butler and Admiral David Farragut steamed up the

Mississippi in the spring of 1862, the rebel soldiers holding Fort Jackson were such unwilling conscripts that they spiked their cannons and shot the officers who refused to surrender.

In **Texas**, opponents of secession included Governor Sam Houston, the man who had led Texas since it became an independent republic following the Mexican War. "Our people are going to war to perpetuate slavery, and the first shot fired in the war will be the (death) knell of slavery," Houston warned.

In Richmond, **Virginia,** a wealthy socialite headed a spy network that funneled military intelligence to a grateful general Grant; some of the information was gathered by one of her freed slaves whom she planted inside the Confederate White House as a dining room servant.

In **Arkansas**, Brig. Gen. Edward W. Gantt was captured with 7,000 other rebel soldiers in April 1862 and, released in August, became disillusioned with the Confederacy and the conduct of the war. After meeting with President Lincoln in Washington, he began a tour of the North, urging citizens to persevere in the war effort.

In **North Carolina**, the 10,000-strong Heroes of America--among the Southern Unionist organizations with secret passwords, handshakes, and oaths—protected deserters and escaped prisoners of war, spied for federal forces, and engaged in sabotage. Its ranks included a husband and wife team who organized a ruthless guerilla band whose memory evokes controversy to this day.

Tennessee Unionist Lucy Williams rode her horse through heavy rain at night to tip off federal forces that Confederate raider and general John Hunt Morgan was spending the night at the home of her mother-in-law in Greeneville.

SOUTH CAROLINA

"South Carolina is too small to be a republic, and too large to be an insane asylum."
James L. Petigru, prominent Charleston attorney

During the first year of the war after South Carolina fired the first shot at Fort Sumter, no active Unionists or disaffection with the Confederacy was evident in the state that was first to secede without a dissenting vote on Dec. 20, 1860.

Prominent Charleston attorney James L. Petigru, a strong opponent of nullification in the early 1830s and of secession in 1860, was a voice crying in the wilderness when he remarked that "South Carolina is too small to be a republic, and too large to be an insane asylum."

By the spring of 1862, however, disaffection was actually reported among the troops garrisoning the Confederate cradle of Fort Sumter. One soldier was court-martialed for using seditious language in the presence of a Charleston civilian after which five of the garrison deserted and some of the cannons were sabotaged. Charges of disloyalty at Fort Sumter continued throughout the following year.

James L. Petigru, Charleston jurist and legislator, opposed South Carolina nullification and secession. Public Domain.

Secessionist fire-eater Robert Barnwell Rhett, editor of the *Charleston Mercury*, was originally a strong advocate of the draft, but by late 1862 he bitterly denounced Confederate conscription as a violation of state rights. "The essence of Conscription is the right to take away the fighting men of the States against the will of both the citizens and the States. It is the right, make what you will of it, to coerce sovereign states," Rhett wrote.

By 1863, many soldiers in South Carolina had grown war weary as their families suffered economic hardship and as hopes dwindled for the Southern cause. Many in the army deserted and returned home while others dodged military service. Some prominent citizens, like those in other rebel states, publicly urged that the war be stopped.

By the summer of 1863, bands of deserters and draft resisters had taken refuge in the mountainous northwestern counties of Greenville, Pickens, and Spartanburg. Operating as groups of between 10 and 30 men, they chased off conscript companies, raided supply depots, and looted the property of pro-Confederates. They built fortified positions on an island in the Broad River and at Jones Gap, Hogback Mountain, Table Rock, Caesars Head, and Potts Camp.

Throughout the Upcountry, men abandoned their homes to hide from the conscription officers and sheriffs who hunted them with bloodhounds, as runaway slaves had been hunted. Some found refuge in the same mountain caves that had sheltered fugitive slaves. Their wives and children formed networks to signal the alarm at the approach of danger.

These well-organized bands, for the most part, wanted only to avoid military service and few regarded themselves as loyal Unionists. South Carolina, after all, was the only Confederate state from which there were an insufficient number of white enlistees to form a Union battalion. However, some 5,000 blacks, mostly former slaves, were enlisted in the federal army.

The activities of deserters and draft dodgers in the northwestern section along the North Carolina border became so troublesome by late 1863 that Gov. Milledge Bonham dispatched troops to the region.

In early 1864, three deserters crossed into Union lines at Morris Island in Charleston Harbor and gave the commander a full description of conditions in Charleston and of the disaffection among Confederate troops on nearby James Island.

In Charleston, the secret Union Association, numbering some 1,400 men and women, helped federal soldiers to escape prison and return to Union lines.

In the Lowcountry southeastern corner of the state at Beaufort, which federal forces had controlled since the Battle of Port Royal in November 1861, a state convention was held on April 17, 1864. The call to the convention invited South Carolinians, without regard to race, to meet and elect delegates to the 1864 Republican presidential nominating convention in Baltimore. About 250 delegates, some two-thirds of them black, assembled and selected 12 whites and four blacks as delegates to the convention.

MISSISSIPPI

*So weakened was Confederate authority that the
county became known as the Free State of Jones.*

At a special Mississippi convention called to
decide the question of secession at the beginning of
1861, delegates voted on Jan. 9 to leave the Union by
an overwhelming 84-15 vote, making the state the
second to secede.

However, following the much-resented
Confederate conscription act in April 1862, a Peace
Society was organized in the northeastern counties,
most of whose delegates to the special convention
had been among the small minority to vote against
secession. The "twenty Negro" exemption of slave
owners enacted later that year fanned the flames of

disaffection, as did exorbitant prices of food and other necessities.

A Peace Society, which included women, met secretly to devise ways of evading the draft. Unionists were especially numerous in Tishomingo, Tippah, Pontotoc, and Itawamba counties where a secret password was "Liberty and union, now and forever." Confederate cavalry and self-appointed vigilance committees ferreted out and punished anti-Confederates, many of whom were jailed, hanged, shot, or tortured to death.

Following the evacuation of Corinth in late May 1862, northern Mississippi was open to the federal army, and Unionists increased their efforts to undermine the Confederacy, including the risky business of spying. Organizing against conscription spread to the southeastern part of the state, especially in Jones, Jasper, Harrison, Jackson, and Hancock counties where a band of Unionists regularly raided the property of loyal Confederates from their outpost on an island in the Leaf River.

In Jones County, dirt farmer Newton Knight, angry because wealthy slave owners could avoid the draft and because Confederate cavalry had seized his horse under the impressment law, deserted from

Confederate ranks and returned home to "fight for the rights and freedom of Jones County."

Knight led a band of some 500 men who drove off a Confederate cavalry unit and conscription agents, ambushed army patrols, and looted warehouses, distributing their stores of food to the needy. So weakened was Confederate authority that the county became known as the Free State of Jones.

Newton Knight's guerillas established "the free state of Jones." Public Domain.

In the spring of 1864, a Confederate regiment swept into Jones, Newton, Covington, and Smith counties to capture deserters—hunting them down with dogs, hanging several, including one of Knight's brothers, and sending some 350 men back to their units.

The native First Mississippi Mounted Rifles, a 600-man Union battalion organized in 1863, aided the federal advance into northern Mississippi, helped beat back Gen. Nathan Bedford Forrest's April 1864 attack on Memphis, Tenn., and later that year prevented Forrest from disrupting Sherman's advance toward Atlanta.

Led by Col. Benjamin Grierson , the First Mississippi joined Illinois and Iowa units in the famous Grierson's Raid through the heart of Mississippi in the spring of 1863. In the movie "The Horse Soldiers" with John Wayne and William Holden, the Southern-accented men, including Ken Curtis, represented the First Mississippi.

The U.S. provost marshal general credited the state with 545 white Union soldiers.

Most, but not all, Native Americans in the South tended to favor the Union when given the opportunity. In the spring of 1863, Eastern Choctaws drafted into the First Choctaw Battalion, Mississippi Cavalry, deserted en masse to the federals as Grant's Vicksburg campaign was getting under way.

The fall of Vicksburg on July 4, 1863, spurred desertions from the ranks. That summer, a conscription officer estimated the number of deserters and stragglers in the state at 5,000. By the end of 1864, the number of deserters was estimated at 7,000. That summer, Gov. Charles Clark told the legislature, "Deserters, thieves and robbers banded together, overawed the citizens."

In September 1864, Gov. Clark defied President Jefferson Davis, a Mississippian, who had called for more men to be drafted for the Confederate war effort. Instead, Clark called for volunteers to defend their home state from Union advances. As the Confederacy crumbled, other state governors held back forces in order to defend their home territory rather than send them to the front.

On July 21, 1864, a group that included prominent citizens from several towns met in Jackson to discuss peace based on a return to the Union. Had the war not ended in April 1865, the fall elections that year might have produced a victory for the peace advocates.

FLORIDA

Several counties in the Panhandle and the southwest area of the state became havens for Florida deserters as well as deserters from other Confederate states.

Delegates to the state convention that assembled in Tallahassee on Jan. 3, 1861 were divided between "cooperationists" who wanted to delay secession until several other states were ready to leave the Union together, and "fire-eater" radicals who called for immediate withdrawal from the United States. The radicals won the debate and the delegates on Jan. 10 voted 62-7 to secede, making Florida the third state to rebel after South Carolina and Mississippi.

As the war progressed, anti-war and pro-Union sentiment and activities increased as a result of conscription, impressments, and federal army advances along the coasts. Some Unionists, including plantation slaves, fled to coastal towns occupied by federal troops. The federals employed many of these escaped slaves on naval vessels and received more than a thousand of them into service as soldiers and sailors.

Florida Unionists generally belonged in one of two classes: Germans and other foreigners and relatively prosperous northerners who had recently settled in seaport towns; and poor backwoods whites who mostly wanted to be left alone to tend their small farms.

Several counties in the Panhandle and the southwest area of the state became havens for Florida deserters as well as deserters from other Confederate states. Some deserter bands, including the 35-member Independent Union Rangers of Taylor County, emerged from the woods and swamps to attack rebel patrols, raid plantations, confiscate slaves, steal livestock, and provide intelligence to Union army units and naval blockaders.

The Independent Union Rangers were headed by William Strickland, who joined a Confederate cavalry regiment when the war broke out. In December 1862, he went absent without leave to his home nearby to care for his sick wife after he was denied a furlough. When he returned to his regiment in the spring, he was court-martialed as a deserter. He deserted again and joined with other deserters and draft evaders living on Snyder's Island in the Gulf Coast marshlands of Taylor County.

Unable to mount an effective assault on the Rangers through the marshes, Confederate soldiers rounded up their families and imprisoned them at Camp Smith near Tallahassee. Angered by the arrests, the Rangers drew up an organizational constitution, established an alliance with the Union naval squadron off the coast, which supplied them with food and arms, and waged a two-year war against the Confederacy.

The best allies of Strickland and other guerillas were local slaves who hid travelling deserters and draft evaders, funneled them food and supplies from the plantations, acted as spies, and often joined them in their raids.

Florida Gov. John Milton appealed to President
Jefferson Davis for help in cleaning out the guerillas
from Taylor, Lafayette, and counties to the south. In
the spring of 1864, anti-Confederates in Calhoun
County just west of the capital devised a plot to
kidnap Milton and turn him over to federal
blockaders off the coast, but he was warned of the
plot and stayed in Tallahassee to avoid capture.

**Florida Independent Rangers skirmish with
Confederate cavalry. Public Domain.**

Gov. Milton believed the impressment law was
largely to blame for the mounting disaffection.
When 52 men from "the best drilled and most
reliable company in West Florida deserted with their
arms, (and) some of whom joined the enemy," the

governor said "they were indignant at the heartless treatment of the rights of citizens."

Other deserters and Unionists totaling about 1,300 joined regular federal units. The First Florida (US) Cavalry was organized in Pensacola in December 1863 and operated in western Florida and southern Alabama. The Second Florida (US) Cavalry was organized mainly from enlistments at Key West and joined federal operations along Florida's western Gulf Coast. Both units took part in numerous skirmishes and raids on plantations and Confederate supplies during 1864 and 1865.

At Jacksonville, which had been occupied and abandoned three times by Union troops, much to the consternation of the local Unionist population, a militia was organized for defense and garrison duty.

Unionist newspapers in Jacksonville, Fernandina, and other occupied coastal towns fanned the embers of discontent regarding the Confederate policies of conscription and impressment of crops, livestock, and slaves for the military. Small farmers, in Florida and throughout the South, resented the draft's exemption of planters and overseers on plantations with 20 or more slaves and its substitution provision allowing wealthier men to pay poorer men to serve for them.

In May 1864, a Unionist convention of delegates from several eastern counties met in Jacksonville and proclaimed their loyalty to the Union.

The only large engagement fought in the state was the Battle of Olustee near Lake City in northern Florida. The February 1864 battle resulted from an attempt by the federal army to force Florida back into the Union prior to the presidential and congressional elections of 1864. However, the federals were forced to retreat in defeat. The rebel victory at Olustee, however, did nothing to win the favor of the local populace, and soon Confederate troops were sent against deserters and other disaffected in Taylor and Lafayette counties.

As the war dragged on and shortages of food and other basic goods increased, war weariness and defeatism grew. On April 1, 1865, eight days before Gen. Lee's surrender at Appomattox, Gov. Milton, unable to bear the impending defeat, took his own life at his plantation home in Jackson County. On May 10, Maj. Gen. Samuel Jones surrendered all Confederate forces in Florida.

At the end of the war, the U.S. Army provost marshal credited the state with 1,290 white Union troops, including six companies of the First Florida Cavalry and five companies of the Second Florida Cavalry. The figure does not include more than a thousand black enlistees, white militia units, or a relatively small number who joined Northern regiments.

ALABAMA

In March 1864, Jackson County in northeastern Alabama seceded from the Confederacy and announced its allegiance to the Union.

In mountainous northern Alabama, where slaveholders were few and resentment of the rich cotton plantation class to the south was strong, a majority had opposed secession in 1861. Before the end of the war, deserters could find refuge there without fear of arrest.

At the state convention in January 1861, the great majority of the 46 percent of the delegates voting against secession were from the northern part of the state. Some proposed that the region form its own state and ask for admission to the Union, and others proposed annexation by Tennessee, which initially had voted against secession.

In July 1861, the governor was informed by a militia captain that a large number of the inhabitants of the counties of Winston, Marion, Fayette, and some of Walker and Morgan, were organizing and equipping themselves to support the federal government. Some of the armed bands fought off Confederate authorities sent to arrest or to draft them. Other bands ventured south into the Black Belt to attack planters and their property—burning cotton stores and cotton gins.

In Alabama and throughout the South, anti-Confederate activists established networks of safe houses and guides that shuttled to federal lines draft dodgers, deserters, and men wishing to enlist in the Union army.

In the state capital and original Confederate capital of Montgomery, an underground of some 30 anti-Confederates kept in close communication, and one Unionist took the risk of providing military intelligence to nearby federal forces, showing how they might attack and occupy the city.

Likewise, Southern blacks took risks to undermine the Confederate war effort, giving refuge to deserters, bringing food to deserter gangs and fugitive slaves heading for Union lines, and spying for federal troops. Some 3,000 former slaves from Alabama became Union soldiers. They built fortifications, guarded bridges and entrenchments, and engaged in combat.

A secret Peace Society--complete with signs, passwords, grips, and oaths-- was organized and soon became strong and active, especially after the Confederate Congress began discussing and then enacting a conscription law in April 1862. Such secret societies sprung up throughout the South.

In April of that year, a convention was held in which delegates from three northern Alabama counties voted to remain neutral in the war, which was widely interpreted that they would join the federal troops when the opportunity came. From these counties, men bearing the Stars and Stripes crossed into northern Mississippi to stir up Unionists there.

The capture of Fort Donelson in February 1862 and the Confederate retreat to Corinth brought federal troops into northern Alabama, where they occupied Huntsville and other towns. This spurred the Peace Society and other Unionists to become more active. They fingered prominent secessionists for imprisonment, acted as agents for the federals to buy cotton, and held pro-Union meetings. The conscription law, widely resented as an infringement on personal liberty, became almost impossible to enforce. One regiment of "hillbilly" and "sand-mountain" draftees deserted as a body.

Resentment of the conscription law contributed to a situation in the mountain region which "menaces the existence of the Confederacy as fatally as the armies of the United States," wrote John A. Campbell, the Confederacy's assistant secretary of war. Pro-Union raiders destroyed Confederate stores, railroads, bridges, and telegraph lines, and acted as guides for federal troops.

A Union captain, following the capture of Florence, Ala., reported, "We have met the most gratifying proofs of loyalty everywhere, across the Tennessee, and in North Mississippi and North Alabama," the welcome being "genuine and heartfelt. Tears flowed down the cheeks of the men as well as of the women."

The "twenty-negro law" that exempted affluent slave owners from the draft caused deep resentment of what some called "a rich man's war and a poor man's fight." A month after Fort Sumter, William Brooks, a leading secessionist, wrote a worried letter to Jefferson Davis that some Alabamians had openly declared that they would "fight for no rich man's slaves."

In the summer of 1863, Confederate general Gideon Pillow was ordered to track down and arrest the state's growing number of deserters. Pillow reported that there were as many as 10,000 deserters in the mountains of north Alabama, and his efforts to capture them yielded only mixed results.

Anti-Confederates were also active in the southeast counties of Dale, Coffee, and Henry (which included present-day Houston and Geneva counties). Guerillas led by Unionist John Ward operated freely during the last two years of the war, finding refuge in the Wiregrass region's pine forests, and sometimes working with federal forces based in Pensacola, Fla. Gov. Thomas Watts called the region, where some guerilla bands were interracial, "the common retreat of deserters from our army, tories, and runaway negroes."

In an August 1863 election, Watts defeated in a landslide pro-Jefferson Davis incumbent John G. Shorter, and the Peace Society elected several of its members to the state legislature and several others to the Confederate Congress.

In March 1864, Jackson County in northeastern Alabama seceded from the Confederacy and announced its allegiance to the Union. That same month, a convention in the federal-occupied city of Huntsville was called to seek readmission into the United States.

In the state capital of Montgomery, several hundred Union soldiers were imprisoned in a warehouse following the April 1862 Battle of Shiloh. Massachusetts-born doctor Edmond Fowler and his wife Martha led a chain of Unionists who provided food and medical attention to wounded and malnourished prisoners. Criticized as being too solicitous of the needs of the enemy, the Fowlers were pressured to move back North.

In the fall of 1864, many soldiers from Alabama and other states deserted to the federals before the battles of Missionary Ridge and Lookout Mountain and the Chickamauga campaign, giving the federals important details about the opposing forces. Many of the rebel prisoners taken by Gen. Grant at Vicksburg were Alabamians and numbers of them, after being exchanged, chose to enlist with the federal army.

The First Alabama Cavalry, recruited mostly from poor farmers in the northern part of the state, was praised for its fighting spirit in a spring 1863 raiding expedition in the Tennessee Valley. From April to September 1864, the regiment took part in Gen. Sherman's advance to Atlanta, acting as scouts and rear guards for the supply line.

After the capture of Atlanta, Gen. Sherman selected the First Alabama as his escort in his March to the Sea during which it destroyed a long stretch of the Atlantic and Gulf Railroad, pillaged plantations, and joined the attack on the seaport city of Savannah. In Sherman's march from Savannah up through the Carolinas, the regiment won praise for its exploits from Gen. Sherman and Union cavalry commander Judson Kilpatrick.

In 1976, a plaque honoring the First Alabama, including a list of the regiment's members, was placed at the courthouse of Marion County in northwest Alabama.

By the end of the war, the U.S. provost marshal general credited Alabama with 2,576 white federal troops, many of whom were native to Mississippi,

Georgia, and South Carolina. This figure did not include home guard and independent units that were never mustered into U.S. service. Several hundred other Alabamians joined out-of-state Union regiments.

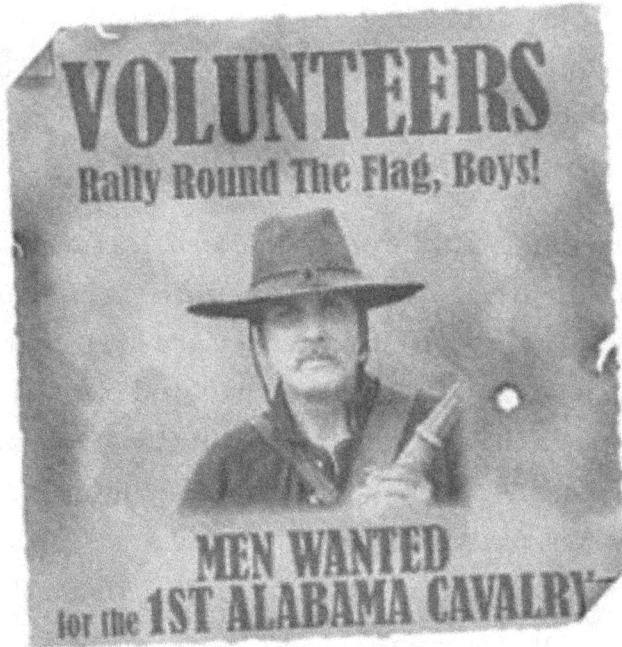

1st Alabama Cavalry recruitment flier

As the prolonged war produced food shortages and sky-high prices throughout the Confederacy, mobs of women from Virginia to Texas looted stores, depots, and warehouses. In Mobile, a small army of women marched through the city with banners reading "Bread and Peace" and looted grocery stores of flour, ham, and other goods. Food riots occurred in several parts of mountainous Randolph County, as desperate women seized stores of wheat and corn. In the southern Alabama town of Greenville, a band of women shouting "Salt or blood" marched on a depot and seized food supplies.

Perhaps the most prominent Alabama Unionist denouncing the Confederacy and urging an end to the war was Jeremiah Clemens, an attorney who commanded a regiment in the Mexican War, served in the state legislature for four years, and entered the U.S. Senate in 1849. As a member of the state secession convention, he opposed secession but did not vote against it.

Early in the war, Clemens commanded a Confederate regiment, but soon had a change of heart and made his way to Philadelphia where he wrote Unionist pamphlets. In an October 1864 appeal to the people of Alabama, he decried the loss of "personal liberty" and "freedom of speech and of the press" as well as the loss of life and property under the "soulless tyranny" of Jefferson Davis.

"If the love of law, order, and tranquility still holds a place in your bosoms—if you are wearied with carnage, and worn out by exactions…you ought to abandon at once the attitude of armed resistance to a Government which never wronged you, and a people whose hearts now bleed in sympathy with yours over the miseries which the mad ambition of your leaders has produced," Clemens wrote.

"Return, as you may do now without dishonor, to the protection of that banner which has been for nearly a century the symbol of freedom and the harbinger of happiness. You have exhibited on the battle-field a heroism which, in a better cause, would have won for you immortal honor. Prove to the world that you are capable of the still higher heroism of daring to do right in defiance of the scoffs or sneers, or threatened coercion of the guilty criminals who led you astray."

Had the war continued until the fall 1865 elections, it is quite possible that the peace organizations would have elected an administration which would have sued for immediate peace and refused further support of the Confederate government.

GEORGIA

Confederate general Gideon Pillow estimated that there were between 25,000 and 30,000 deserters and draft dodgers in the state.

At the 1861 Georgia secession convention, many prominent men, including U.S. Congressman Alexander Stephens, later vice president of the Confederacy, argued that the state and its institution of slavery would fare better within the Union than without. Although 44 percent of the delegates voted against secession, they all signed the Jan. 19, 1861 ordinance of secession in order to present a united front, making Georgia the fifth state to leave the Union.

Soon after secession, a Peace Society was organized in mountainous northwest Georgia and was in contact with Alabama's Peace Society across the state border. Many mountain county courthouses in the northern part of the state refused for a time to lower the Stars and Stripes. A network of pro-Union spies based in Murray County operated throughout north Georgia.

Dissent was not entirely confined to north Georgia. In Crawfordville, in the heart of the state's cotton belt, a crowd of citizens gathered to declare that the election of Lincoln was no cause for disunion.

An Augusta newspaper editor warned, "The greatest danger to the new Confederacy arises not from without, not from the North, but from our own people."

As the war entered its second year, dissension grew and was inflamed by the conscription law and the 10 percent tax-in-kind on agricultural products, which fell heaviest on poor farmers. Remarkably, dissension reached all the way to the top. Vice President Stephens and Georgia Gov. Joseph E. Brown opposed several policies of the Jefferson Davis administration, including conscription and suspension of the writ of habeas corpus, complaining that personal liberty and state rights were being violated.

Gov. Brown was so enamored of the doctrine of state rights that during the rather short interval of complete state sovereignty between Georgia's secession and its joining the Confederacy, he sent a diplomatic representative to Europe to seek recognition for Georgia from Great Britain, France, and Belgium.

Brown gave tacit support to draft evaders and deserters by declaring that the April 1862 conscription law was an unconstitutional injustice to the citizens of Georgia and that he would not cooperate with its implementation. He called the law "a bold and dangerous usurpation by Congress of the reserved rights of the States, and a rapid stride towards military despotism."

Georgia Governor Joseph E. Brown opposed the conscription law and other Confederate policies. Courtesy of Hargrett Rare Book and Manuscript Library. University of Georgia Libraries.

Brown even threatened to disarm Georgia volunteers who left the borders of the state at the summons of the Confederate military. He withheld from the draft more than 8,000 men he said were essential to the functioning of the state government. Other Southern states, led by North Carolina, withheld nearly 22,000 men from military service for the same argument.

In late 1862, the Confederate secretary of war reported that more than half of the army's volunteers from northern Georgia were at home without leave. Most of these men were hiding in the mountains to avoid arrest.

The pine barrens Wiregrass section of southeast Georgia and Okefenokee Swamp bordering Florida were home to some 1,000 deserters and anti-Confederate gangs which fought fierce skirmishes against Confederate forces. In the fall of 1862, in southwest Georgia's Marion County, some 40 pro-Unionists turned a large house into a fortress and refuge. State militia and Confederate cavalry were repeatedly unable to capture the stronghold. In late 1863, an expedition of state and Confederate troops passed through Wilcox, Ware, Coffee, and Clinch counties, arresting deserters and runaway slaves.

By early 1863, disaffection was fed by widespread food shortages, especially in north Georgia, where counties including Rabun, Union, and Gilmer became a hotbed of deserters and Peace Society activists. The ruling planter class was resented throughout the South for devoting prime farmland to cotton instead of food production. Gov. Brown in 1863 lamented that impressments of food and livestock from small farmers "have been ruinous to the people of the northeastern part of the state."

Elsewhere, bands of hungry women in Atlanta, Columbus, and Milledgeville raided food stores. In the 1864 Savannah bread riot, the women rioters cried for "bread or blood" and dispersed only when army troops were called out.

Anti-Confederates banded together to prevent their own arrest and to liberate those who had been arrested. These bands commonly subsisted by plunder as did their counterparts in pro-Union sections of Alabama, North Carolina, and Tennessee. Gov. Brown declared them outlaws and sent the state militia to arrest deserters and ring leaders and break up their organization. Confederate Gen. Gideon Pillow estimated that there were between 25,000 and 30,000 deserters and draft dodgers in the state.

Joshua Hill, a prominent Unionist who served as a U.S. senator prior to secession, garnered nearly a third of the vote for governor in the fall 1863 election in which Brown was reelected. Gen. Sherman spared Madison, where Hill resided, on his path of destruction in 1864. In the same election, voters rejected the incumbents and elected nine new delegates to the Confederate Congress, eight of whom were opponents of Jefferson Davis.

When Gen. Sherman entered northern Georgia in the spring of 1864, hundreds of Unionists from the surrounding hills rushed to join them, and in the town of Jasper, some 1,000 men organized a Unionist home guard.

Sherman's arrival in the state increased dissension among civilians and soldiers and numerous peace meetings were held. When President Davis travelled to Georgia to rally the people to the cause and called on the state militia to aid in defending the state from Sherman's advance toward Atlanta, Gov. Brown countered that the Georgians who were with Gen. Robert E. Lee's Army of Northern Virginia should be returned to Georgia to do the fighting.

As Sherman's March to the Sea advanced toward the seaport of Savannah, Gov. Brown refused a Confederate general's request that slaves be impressed for the construction of fortifications to defend the port city.

Meanwhile, Vice President Stephens, who had been in his home state instead of the Confederate capital of Richmond for two years, declared in Augusta that "the resources of the South were exhausted and that peace ought to be made."

In Atlanta, the Union Circle of about a hundred men and women helped men escape the South to evade military service, acted as spies for invading federal troops, and provided food and medical treatment for prisoners of war. Cyrena Stone was a leader among those who aided the Union prisoners in their squalid living conditions. Stone collected money, food, and medicine for the prisoners and passed it secretly to them during her frequent visits.

Confederate authorities decided to crack down on the Atlanta Unionists aiding federal prisoners and engaging in other disloyal acts ascribed to them. In the late summer of 1862, they declared martial law in Atlanta, giving the provost marshal broad powers to arrest and punish suspected traitors and criminals.

About a dozen suspects, including three women, were arrested and jailed, and one man died after a brutal beating during what was described as a six-month "reign of terror."

Atlanta vigilance committees interrogated those suspected of disloyalty and those who expressed Unionist sentiments were ordered to leave the state, and some were whipped or lynched.

In mountainous Pickens County, more than a hundred men formed a Union home guard to defend themselves against the Confederate home guard and to aid Sherman's advancing federals. In August 1864, the federals recruited about 300 Georgians, some of whom were sent to the Fifth Tennessee Mounted Infantry, and the rest enrolled as the First Georgia State Troops Volunteer Battalion. The First Georgia fought a running guerilla war that occupied Confederate loyalists who otherwise would have been battling the invading federals.

While Sherman was in Savannah, Unionists in nearby Tatnall and Liberty counties met secretly and passed resolutions pledging loyalty to the Union and offering any support to Sherman's occupying army. Sherman replied to the resolutions by offering aid and protection to the loyalists. During Sherman's march, thousands of slaves fled to Union lines, draining plantations of their labor and providing military intelligence to the federals.

In February 1865, in Irwin County in southern Georgia, Unionists and deserters held a meeting where they called for the Confederacy to surrender. When a rebel militia lieutenant tried to break up the meeting, he was knocked down with a musket wielded by a man who led three cheers for Lincoln. Demands for peace continued until the war ended.

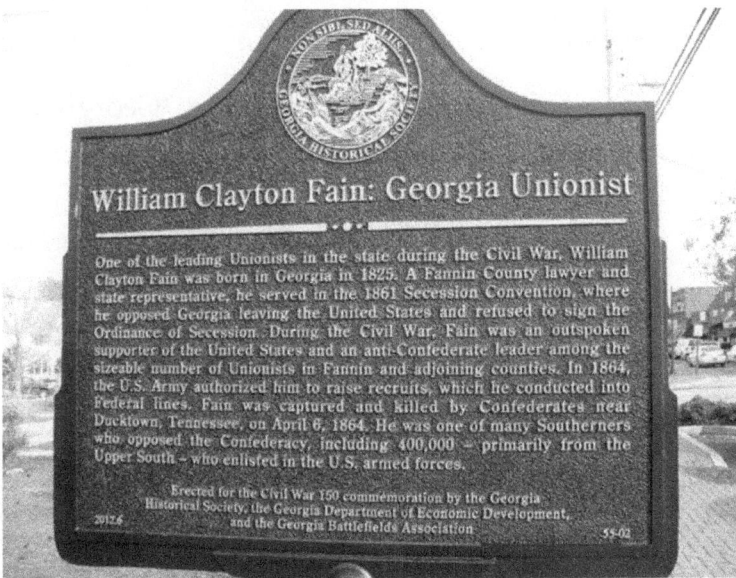

Marker in Blue Ridge, Ga., 90 miles north of Atlanta, honoring slain Unionist William Clayton Fain. Courtesy of David Seibert.

The Georgia Historical Society in October 2012 unveiled an historical marker in Blue Ridge, Ga., with the inscription, "William Clayton Fain: Georgia Unionist." A Fannin County state representative, Fain served in the 1861 secession convention where he spoke against Georgia leaving the Union and refused to sign the Ordinance of Secession. He was a prominent anti-Confederate leader among the sizable number of Unionists in Fannin and adjoining counties. In 1864, the U.S. Army authorized him to raise recruits, which he conducted into federal lines. He was captured and killed by Confederates in April 1864.

Another historical marker honoring "Georgians in the Union Army" was erected in 2013 in the North Georgia county of Dawson.

Approximately 400 white Georgians enlisted in the federal army, fewer than any Southern state except for South Carolina. This included a battalion of some 200 soldiers and at least another 200 to Northern regiments.

Historical marker in Dawson County commemorates Georgians who fought for the Union. Courtesy of David Seibert.

LOUISIANA

*When Maj. Gen. Benjamin Butler and Admiral
David Farragut steamed up the Mississippi in
April 1862, the Confederate soldiers holding Fort
Jackson were such unwilling conscripts that they
spiked their cannons and shot the officers who
refused to surrender.*

In Louisiana, voters in January 1861 approved
secession by only a 54 percent majority, and state
officials delayed releasing the likely fraud-tainted
election results for three months. In the special
convention in Baton Rouge that followed the
referendum, the Jan. 26 vote for secession was 81 to

50, making Louisiana the sixth state to leave the Union.

However, in the southern Cajun parishes of Ascension, Assumption, Terrebonne, Jefferson, Plaquemines, and St. Bernard, the popular vote against the election of secession delegates had been more than 2 to 1, and in St. James and St. John the Baptist parishes, the vote was more than 4 to 1. The mostly non-slaveholding Cajuns held no love for the state's rich planters and Cajuns enlisted in federal military units in considerable numbers.

The largest geographical area of Unionism centered on Caldwell, Catahoula, and Winn parishes in the north central part of the state, which voted more than 2 to 1 against secession and remained strongly loyal to the United States throughout the war, and where there were frequent violent clashes between Unionists and Confederates.

By the summer of 1861 in New Orleans, the war had produced shortages and rampant inflation, and some 10,000 wives of soldiers and other women marched to the mayor's office to demand aid for food and rent.

Unionists faced violent reprisals in some parts of the state, as when Gen. Nathaniel Banks' army withdrew from Alexandria in May 1864 and Unionists' homes were burned and some were killed by Confederate soldiers. In response, some Unionists, called "jayhawkers" in Louisiana and elsewhere in the South, formed guerilla units to protect fellow Unionists, serve as scouts for federal forces, and raid Confederate civilian property.

One jayhawker, Capt. Dennis Haynes of the 1st Louisiana Battalion Scouts, operated for several months in central and western Louisiana and participated in the Red River campaign, a series of battles in the spring of 1864 in which Union forces were defeated. Haynes wrote a postwar account of his adventures, including being shot, titled "A Thrilling Narrative: The Memoire of a Southern Unionist."

When Maj. Gen. Benjamin Butler and Admiral David Farragut steamed up the Mississippi in April 1862, the Confederate soldiers holding Fort Jackson were such unwilling conscripts that they spiked their cannons and shot the officers who refused to surrender. The fort fell without a Union shot being fired.

On May 1, 1862, Butler captured New Orleans and thousands of slaves fled the plantations to the safety of Union camps. Gen. Butler organized three regiments of free blacks and creoles--the 1st Louisiana Native Guards, which superseded a disbanded would-be Confederate unit of free blacks with the same name, but that the Confederates refused to employ. The Native Guards became the first black unit to fight in a major engagement in the war.

Butler won support and valuable intelligence from civilians by distributing food stores to the poor, taxing the rich to provide social services, and expanding the city's sewer system to prevent a recurrence of epidemic yellow fever.

In December 1862, Gov. Thomas Moore warned Jefferson Davis that unless the Confederate government sent back Louisiana troops to defend the state from Union forces, the citizenry might demand that the state secede from the Confederacy.

In the summer of 1863, the 1st Louisiana Native Guards played a key role in an assault on Confederate fortifications at Port Hudson on the Mississippi River, for which they were widely praised for their bravery and coolness under fire. That September, the regiment swept through southern Louisiana, freeing slaves and recruiting many of them to their ranks. The 1st and 2nd

Louisiana Infantry and companies of the 1st Louisiana Cavalry also distinguished themselves in the siege and capture of Port Hudson on July 9, 1863, five days after the fall of Vicksburg.

Gen. Banks, Gen. Butler's successor as military governor of New Orleans, in May 1863 issued an order to organize freedmen into other military units, generally known as the Corps d'Afrique. In addition to fighting, the former slaves built fortifications, roads, dams, and canals, repaired levees, and acted as guards and scouts.

After the U.S. Congress passed the conscription act in March 1863, Gen. Banks imposed conscription on the New Orleans area, drafting hundreds of men into the "New Orleans Volunteers" which was charged with defending the city from Confederate attack.

In the final weeks of the war, the 1st Louisiana Cavalry, moving from Pensacola Bay toward Mobile, led the charge at the victorious Battle of Bluff Springs—Private Thomas Riley of the regiment capturing the battle flag of the 6th Alabama Cavalry and receiving the Medal of Honor.

Unionist activities took rebel soldiers from the battlefield in order to police the home front, thus weakening the Confederate war effort in Louisiana and the other rebel states.

James Madison Wells, the owner of a large cotton plantation near Alexandria in central Louisiana, denounced the Confederacy as a rich man's government. Wells organized bands of guerillas to raid Confederate supply lines and depots and attack home guard units. By 1864, Union troops controlled most of south Louisiana and in March of that year Wells was elected lieutenant governor.

Unionist Michael Hahn was a moderate Republican who had served as state attorney general and as one of two Louisiana members of the U.S. House of Representatives elected March 4, 1861. He befriended President Lincoln during his time in Washington, and was elected governor in 1864 under President Lincoln's 10 Percent plan (the minimum percentage of whites needed to swear an oath of allegiance to the United States and assent to emancipation) for rejoining the Union.

Unionist Michael Hahn served in Congress during the war and took office as Louisiana governor on March 4, 1864. Public Domain.

The number of white Union troops officials credited to the state was 5,224. The figure did not include Louisianans who joined out-of-state regiments, which Gen. Butler estimated at more than 1,200 after the occupation of New Orleans. The figure did not include the uncounted numbers in militia and home guard service or the numbers of free blacks and ex-slaves serving in U.S. regiments.

TEXAS

"Our people are going to war to perpetuate
slavery, and the first shot fired in the war will be
the (death) knell of slavery."

Texas Governor Sam Houston

Texas became to seventh state to secede from
the Union on a 166-7 vote of the legislature on Feb. 1,
1861. Other than South Carolina, where the vote was
unanimous, this was the highest percentage of any
other state.

The decision was affirmed on Feb. 23 when
voters in a statewide referendum approved secession
by a majority of 46,129 to 14,697, or about 75 percent.

Among slaveholders, 81 percent voted for secession while only 32 percent of non-slaveholders did so.

The secessionists, however, did not include the man who had led Texas since it became an independent republic following the Mexican War. "Our people are going to war to perpetuate slavery, and the first shot fired in the war will be the (death) knell of slavery," warned Texas Gov. Sam Houston.

Governor Sam Houston opposed secession. Public Domain.

Calling secession a "rash action," Houston predicted, "After the sacrifice of countless millions of treasure and hundreds of thousands of lives, you may win Southern independence if God be not against you, but I doubt it. The North is determined to preserve this Union. They are not a fiery, impulsive people as we are, but once they begin to move in a given direction, they move with the steady momentum of a mighty avalanche, and what I fear is that they will overwhelm the South with ignoble defeat."

Houston reluctantly accepted secession but urged that Texas revert to its former status as an independent republic and to remain neutral. Refusing to take an oath of allegiance to the Confederacy, Houston was deposed from office on March 16, 1861.

Houston wasn't the only prominent Texan to oppose secession. Andrew Jackson Hamilton was an Austin attorney, a state attorney general, and elected in 1858 to the U.S. Congress, where he spoke out against slavery long before most Southern Unionists came to favor emancipation. He fled to Mexico with a small band of supporters in the spring of 1862. Leaving Mexico the same year, Hamilton made his way to Washington and met with President Lincoln in October 1862. He was commissioned as a brigadier general in the Union army, and Lincoln appointed him military governor of Texas in exile.

Hamilton spent much of the war living in the North and New Orleans, speaking in favor of reunion and urging a military invasion of his home state. Returning to Brownsville with the federal army, he issued a pamphlet on Jan. 1, 1864, urging his fellow Texans to abandon the "Slave Aristocracy" and the "despotic power of the class over the mass" and to support the Union cause for "the restoration of freedom, justice, and order."

In June 1865, President Andrew Johnson appointed him provisional governor of Texas. He supported black suffrage as governor and helped to draft the Texas constitution of 1869. Losing the race for governor in 1869, he retired from public life.

In his "Address to the People of Texas" pamphlet, Hamilton said the Confederacy had utterly failed to deliver on its promise of an easy victory over the North by a more free and prosperous South where King Cotton would reign supreme and bring recognition from foreign governments. However, "Where all was peace, security, and contentment—now all is strife, decay, and wretchedness."

Hamilton said everyday life had been made miserable by the imposition of martial law, travel restrictions, lack of free expression of opinion, impressment of one-tenth of a farmer's crops, worthless Confederate currency, and the unfair exemption of "20 Negro" plantation owners from military service.

Compounding the misery, Hamilton pointed to the depredations of "the murderers and ruffians of the country banded together in secret societies known as Sons of the South…Three thousand of your citizens have perished because they loved good government, and peace, and order in society— perished as felons. They have been hung, shot, and literally butchered—they have been tortured, in many instances, beyond anything known in savage warfare."

As for the underlying cause of the war, Hamilton asked, "What has slavery done for you, that you should become its champion and defend it to the death? It never benefitted you, materially, morally or politically. It has, in fact, been a clog upon your prosperity—the enemy of truth and virtue, and a sore upon the body politic for long years, producing constant irritation."

"I appeal to you," he wrote, "in behalf of your own interests, present and future, to cease a struggle so hopeless and disastrous."

Another defector was Edmund J. Davis, a district judge in the Rio Grande Valley, who refused to take the oath of allegiance to the Confederacy and boarded a ship bound for Washington, D.C. There he convinced President Lincoln that it was possible to reclaim Texas by rallying the state's Unionists.

It is believed that many more Texans could have been rallied to the Union cause and enlisted in the federal army were it not for the failed military campaign to seize Galveston and the failed Red River campaign in western Louisiana. The federal capture of Brownsville on the state's southern tip in October 1863 gave impetus and opportunity for federal recruitment. The recruits included 33 blacks who joined the Corps d'Afrique.

Many pro-Unionists also were found in some counties of north-central Texas. The latter region was settled mostly by people from the states of the Upper South, where secession had been rejected until the firing on Fort Sumter and Lincoln's call for troops to put down the rebellion.

Sherman, Texas, newspaper editor E. Junius Foster called for North Texas to secede from the state and stay in the Union as a free state. Foster was murdered in 1863 after he applauded the assassination of a Confederate colonel who had arrested pro-Unionists who were put on trial and hanged.

In the southwestern part of the state, Col. Charles Anderson, brother of Col. Robert Anderson of Fort Sumter fame and leader of the Unionist opposition in San Antonio, was arrested in October 1861.

In December 1862, a group of draftees in Austin County assembled and assaulted their would-be captain, driving him off with sticks and iron bars. A week later, some 600 men from Austin, Washington, Fayette, Lavaca, and Colorado counties, many of them Germans, held a meeting and decided to organize themselves into companies of infantry and cavalry for mutual defense against service in the Confederate military. In 1863, about 2,000 deserters fortified themselves near the Red River in defiance of the Confederacy.

East Texas gave much support to secession, and the only counties in that part of the state in which significant numbers of people opposed secession were Angelina, Fannin, and Lamar.

Confederate conscription was generally unpopular throughout the South, especially so because it exempted men who owned 20 or more slaves, causing many to argue that the war was being fought by poor people on behalf of the wealthy minority. Some Texans hid out in the hills and woods to avoid conscription while others fled the state, some joining the Union army.

In the 1863 election for governor, Pendleton Murray, a vocal opponent of Jefferson Davis, defeated pro-Davis candidate Thomas J. Chalmers by more than 5,000 votes. Murray balked at Confederate conscription of troops from the Texas militia on the ground of state rights and home defense.

On May 30, 1862, to clamp down on Unionist activity, a Confederate general put Texas under martial law, with travel restrictions and other unpopular regulations.

The largest concentration of anti-secessionist sentiment was among the immigrant German population of some 20,000 in the Texas Hill Country of central and south Texas, where Confederates fought and massacred a band of German Texans along the Nueces River in August 1862.

The war was especially unpopular among these non-slaveholding Germans who had fled to America to escape military service and the turmoil of the 1848 European revolutions. They also were anti-slavery and highly patriotic.

In the summer of 1861, some 500 of these German immigrants formed the Union Loyal League with the aim of restoring the federal government in Texas. The Germans also formed the Hill Country Militia with the aim of resisting the Confederate loyalty oath and the draft, and supporting the federal army when it entered Texas.

To escape Texas and the martial law crackdown, many of the Germans fled across the Rio Grande to Mexico. In August 1862, a group of families totaling 61 individuals assembled west of Kerrville and headed for the border, some with the intention of making their way to Union-occupied New Orleans and joining the federal army. Confederate cavalry pursued them for seven days and a hundred miles.

On Aug. 10, just a day's ride from Mexico, the Confederates ambushed their camp in Kinney

County near the Nueces River. In the skirmish that ensued, about half the 61 Germans were killed, nine of them after being wounded and taken prisoner. Of those who escaped the Nueces Massacre, nine more were later shot trying to cross the Rio Grande. Among the 96-man Confederate company, two were killed, including its commander, and 18 wounded.

A few who escaped made their way to New Orleans and joined the federal First Texas Cavalry. But the incident generally put an end to active Unionism among the Texas Germans for the remainder of the war.

On Aug. 10, 1866, to commemorate those who died in the Nueces Massacre, a German-language Treue der Union monument was dedicated in Comfort, Texas. The remains of those who were killed are buried at the site of the monument. Today, few of the Texas counties settled by Germans display the ubiquitous Confederate infantrymen on the courthouse square.

Monument commemorating the Nueces Massacre was dedicated in 1866 in Comfort, Texas. Public domain.

An incident of vigilante justice known as the Great Hanging occurred in largely anti-Confederate Cooke County in north Texas in October 1862. About 150 men who had failed to report for the draft, some of whom were alleged to belong to the secret Union League, were arrested by Confederate cavalry.

Accused of plotting to seize two local arsenals and of planning to cooperate with Union forces poised to enter northern Texas, a court was convened in Gainesville to try the men for treason and draft evasion.

During jury deliberations in Gainesville, a mob stormed the courtroom and demanded a list of the men on trial; 14 of the men were hanged the next day. Meanwhile, the colonel who commanded the cavalry regiment which had arrested the alleged conspirators was ambushed and killed. This prompted the jury to reconvene and 19 additional suspects were convicted and hanged.

Confederate President Jefferson Davis condemned the hangings and dismissed the state military commander who had put all of Texas under martial law in defiance of Davis' wishes.

A North Texas Confederate company serving in Arkansas almost mutinied when they heard about the mass hangings, and some deserted. Fear of arrest and lynch mobs caused hundreds of Unionist families to flee North Texas.

A granite monument in a small park in Gainesville marks the spot of the Great Hanging.

By early 1865, desertion was rampant in every part of the state as it became obvious that the Confederate cause was lost.

The U.S. provost marshal credited Texas with 1,965 white troops, including Hispanics, organized in two regiments of cavalry, one company of partisan rangers, and one company of scouts. Not included in the total were those men who enlisted in units from other states, including those who made their way to Arkansas.

VIRGINIA

*Elizabeth Van Lew's couriers smuggled out
military intelligence in empty egg shells and even
brought flowers from her garden as gifts to
general Ulysses S. Grant.*

The Old Dominion state was late in joining the
Confederacy, initially voting against secession, and
preceding only North Carolina, Arkansas, and
Tennessee in leaving the Union. Strong Unionist
sentiment split the state in two, giving birth to the
new free state of West Virginia.

Anti-Confederates flourished in the Union-
occupied areas of Northern Virginia near the District

of Columbia and were especially active in Southwest Virginia where they fought guerilla actions which diverted Confederate troops from opposing federal forces on the battlefield. And an underground Unionist spy ring operated in the state and Confederate capital of Richmond.

Delegates to the Virginia Convention in Richmond on April 4, 1861, voted 90 to 45 against secession. However, after the Battle of Fort Sumter and Lincoln's call for troops to put down the rebellion, the convention reversed itself, approving the Ordinance of Secession on April 17 by a vote of 88 to 55.

With nearly 40 percent of the delegates voting to remain in the Union, the convention agreed to put the question to a popular referendum on May 23. As the date approached, Unionist sentiment waned as Virginia initiated military action to capture the federal arsenal at Harpers Ferry and offered Richmond as the new capital of the Confederacy. Newspapers beat the war drums and called a vote against secession an act of treason.

The referendum's official vote count was 125,950 to 20,373, or more than 86 percent, in favor of secession, but the balloting was not secret and

Unionists charged widespread intimidation, including death threats. In Rockingham County in the Shenandoah Valley, secessionist vigilantes forced those who had cast Union ballots back to the polls at gunpoint to change their votes. Unofficial tallies, which probably were more accurate, showed that loyalists polled about 20 percent as opposed to less than 14 percent in the vote certified by the governor.

The secession crisis was so heated and bitterly divisive that the *Richmond Examiner* admitted that the South "was more rife with treason to her own independence and honor than any community that ever engaged before in a struggle with an adversary."

At the Virginia Convention on April 17, the state's western delegates voted against secession by a margin of nearly 2 to 1, laying the foundation for the western counties to secede from Virginia and establish the state of West Virginia in 1863, thus depriving the Confederacy of population and resources. The western counties had close economic ties to Ohio and Pennsylvania and, with relatively few slaves, resented the political influence of the state's eastern counties whose voting strength in the state legislature was bolstered by counting their slave population.

By the summer of 1861, the U.S. Army under generals George B. McClellan and Jacob D. Cox, aided by a pro-Unionist population, had gained control of most of northwestern Virginia, including Charleston. Union successes encouraged an increasing number of Virginians to fight for the Union cause, starting with the First Regiment of Virginia Infantry, a U.S.-armed militia recruited from the Wheeling area in the state's northern panhandle.

In June 1861, delegates to a convention set up a Restored state government, with its capital in Wheeling, and began to take steps to establish the separate state of West Virginia. Its governor was wealthy lawyer and merchant Francis H. Pierpoint, who began to recruit a large number of volunteers for the federal army and the state militia. As of June 30, 1862, a total of 11,428 volunteers from "Virginia (Western)" had joined the U.S. Army, according to the U.S. adjutant general.

Volunteering early in the war, the First Virginia Infantry and four companies of the Second Virginia Infantry, dispatched by Maj. Gen. McClellan along with units from Ohio, routed rebel forces at Philippi on June 3, 1861, to clear them from western Virginia. Among all the Virginia/West Virginia Union regiments which did good service during the war, perhaps the Seventh Virginia Infantry was the most distinguished, having fought, in chronological order, at Romney, Antietam, Fredericksburg, Chancellorsville, Gettysburg, Bristol Station, Mine Run, Wilderness, Spotsylvania, Totopotomoy, North Anna, Cold Harbor, Petersburg, Strawberry Plains, Deep Bottom, Reams' Station, Boydton Road, and Appomattox.

After a county by county vote, West Virginia became a state on June 20, 1863, with a provision requiring gradual emancipation. Leaving the capital of the new state in Wheeling, Pierpoint moved the Virginia government-in-exile to Alexandria, across the Potomac from Washington, from which he governed military-occupied Northern Virginia, the Eastern Shore, and the Norfolk area.

The philosophical basis for West Virginia statehood was expounded by Whitman T. Willey, a Morgantown, Va. (later part of West Virginia) attorney, who was a delegate to the 1861 Virginia secession convention. He voted against secession, and was elected to the U.S. Senate in 1861 as a member from the Restored government. He became a leader of the West Virginia statehood movement and in 1863 was elected to represent the new state in the Senate.

In an 1862 speech, Willey said, "Our Federal Constitution is a farce if this right of secession be admitted." He said the South, which "has always exercised a controlling influence in the councils of the Republic" and could have maintained slavery within the Union even after the war began, had neither the right nor any legitimate cause to secede. "No law could have been enacted which was not acceptable to the South," he said.

Secession, he said, "was no sudden, unpremeditated insurrection—it was the result of a deliberate, long-conceived conspiracy, as has been virtually acknowledged by many of the principal conspirators." The underlying cause was not the false rallying cry of "Southern rights," but rather "dissatisfaction with the principles and operation of democratic government." It was the slaveholding planter aristocracy's "hostility to the simplicity and equality of republican institutions," he argued.

From the western portion of Loudoun County in Northern Virginia arose the Loudoun Independent Rangers, recruited by prominent county businessman Samuel C. Means. The Rangers scouted for federal troops and engaged with mixed success in operations in Loudoun County against Confederate cavalry based in the eastern part of the county. The

Rangers kept John Mosby's cavalry busy, drawing it away from Washington, D.C., and served with Union forces during the successful Shenandoah Valley Campaign of 1864.

Attempts to recruit for the army in eastern Virginia netted only enough men to fill a company of the First Loyal East Virginia Infantry. Other eastern Virginians, along the Potomac below Alexandria, belonged to the Accotink Home Guards.

More than 5,700 Virginia blacks, ex-slaves and free, were mustered into service with the United States Colored Troops. Slaves who fled the plantation to Union lines often labored for the army as teamsters, guides, and cooks. And escaped slaves drained the Confederacy's pool of labor. Those remaining as slaves assisted the federals by providing military intelligence and helping captured Union soldiers to escape to Union lines.

Although the planter aristocracy led the movement for secession in Virginia and the other Southern states, they balked at military commanders' requests to lend their slaves to labor for the common cause, such as by building military fortifications. Thus, for example, Confederate fortifications around

Yorktown in preparation for McClellan's advance in early 1862 were left incomplete.

Although a majority of the population of Southwest Virginia voted for secession in 1861, anti-Confederate sentiment and activity increased there during the war, sparked largely by the Confederate Conscription Act passed on April 16, 1862, the first military draft in American history.

The mountainous region, where slaveholders were few, became a haven for draft dodgers and deserters, and a stronghold of the Order of the Heroes of America, a secret Unionist society based in western and central North Carolina, and stretching into East Tennessee. Heroes of America members provided intelligence to the Union army, assisted draft dodgers and deserters from the Confederate army, and participated in armed bands that plundered Confederate property for their subsistence.

Heroes of America had the grip, signs, and passwords which members used to approach and communicate with each other. The federals allowed members of the organization to pass freely through their lines. To induce people to join the society, the federals offered protection of their persons and property and, after the war, a share in the real estate of those who had been loyal to the Confederacy.

Unionist activity in Floyd, Montgomery and other southwestern counties prompted the Confederate government to send troops to the area to supplement the local home guard and to carry out courts-martial and executions, but the effort met little success in capturing or controlling runaway soldiers.

In the fall of 1864, anti-Confederates centered in Floyd and Montgomery counties organized what they called the State of Southwest Virginia for which they elected a governor and other officials. The same counties organized a brigade of deserters headed by a leader designated as "general."

In addition to geography, Virginia Unionism was also influenced by religion and nationality. The Shenandoah Valley contained many settlers of German origin who were members of pacifist sects – Mennonites, Amish, and Dunkards—who had come to America largely to escape military service in European wars and who also disapproved of slavery. Resistance to the draft by members of the pacifist sects was so troublesome that in June 1864 the Confederate government exempted them following months of persecution.

Antiwar Mennonites in the Shenandoah Valley shuttled hundreds of deserters and draft refugees out of the Confederacy, using an underground railroad of churches and homes as safe houses.

In the Army of Northern Virginia, desertion was so common that as early as 1862, generals Robert E. Lee and "Stonewall" Jackson began frequent executions.

During the spring of 1863, food riots broke out across the South, the largest occurring in Richmond. The war left wives and widows to fend for themselves, and as planters grew too much cotton and tobacco at the expense of food crops, food prices skyrocketed. After Richmond's city council refused a slight increase in poor relief, hundreds of women on April 2 assembled at Capitol Square and headed toward the business district.

Their ranks growing to more than a thousand as they marched, they smashed down shop doors and looted stock. Gov. John Letcher threatened to shoot the rioters, who continued their rampage as fire hoses were turned on them. President Davis mounted a cart and threw money at the mob, but threatened to order them shot if they did not disburse. As they scurried off with their loot, the police arrested a few stragglers.

The most secretive and effective den of Virginia Unionism was located in Richmond, right under the nose of the Confederate government. What one historian called "the most productive espionage operation of the Civil War" was headed by high society socialite Elizabeth Van Lew, a remarkable woman who defied the conventions of the 19th century South. Confederate authorities suspected her, but could not believe a proper and seemingly eccentric unmarried lady called "Crazy Bet" could be a spy even as she brought together Union sympathizers from all levels of society, from slaves to slaveholders.

The spy network made contacts with the federal army, including Maj. Gen. Benjamin F. Butler, and assisted 59 Union officers in escaping from Richmond's Libby Prison. Van Lew often visited the prison, bringing food, medicine, and clothing bought with her own money. She collected information from the prisoners about Confederate troop movements they observed after bring captured. To deflect suspicion from herself, she played her Crazy Bet role, wearing unusual clothing and mumbling to herself as she walked the streets of Richmond.

The network extended into the Confederate White House where one of Van Lew's freed slaves, Mary Elizabeth Bowser, worked as a dining room servant. Bowser funneled anything worthy of note to Van Lew, whose elaborate courier system passed on the information to the federals at City Point (now Hopewell) or Fortress Monroe in Hampton. On several occasions, Bowser was present when Davis met with Robert E. Lee and members of the Confederate cabinet.

The Van Lew mansion on Church Hill contained a secret room where she hid escaping Union soldiers and runaway slaves before moving them through a network of safe houses which she established. Van Lew and her mother and brother entertained prominent Richmond citizens and various Confederate officers and government officials from whom Van Lew picked up pieces of military intelligence. Van Lew's couriers smuggled out military intelligence in empty egg shells and even sent flowers from her garden as gifts to Lt. Gen. Ulysses S. Grant.

When Lee evacuated Richmond, Grant sent Union cavalry into the city to protect Van Lew, who had been threatened by a mob after flying the American flag from her rooftop. The Union soldiers found her inside the Confederate War Department, collecting documents she thought might be useful to Grant. After visiting Van Lew in her mansion, Grant wrote her, stating, "You have sent me the most valuable information sent from Richmond during the war."

Richmond socialite Elizabeth Van Lew ran an elaborate spy ring. Public domain.

The war left Van Lew virtually destitute and a social outcast in Richmond society. When Grant became president in 1869, he appointed her postmaster of Richmond.She died in 1900 and was

buried in Shacke Cemetery where a large stone, a gift
of the Revere family in Massachusetts, carries the
inscription: "Elizabeth L. Van Lew, 1818-1900: She
risked everything that is dear to man—friends,
fortune, comfort, health, life itself—all for the desire
of her heart—that slavery might be abolished and the
Union preserved."

ARKANSAS

"And, after nearly two years of strife, we awaken
from a fearful baptism of blood to the terrible
truth that the shadow of the despotism which we
fled from under Mr. Lincoln dissolves into
nothingness compared to the awful reign of
tyranny that we have groaned under at the hands
of Jefferson Davis and his minions."
Edward W. Gantt, former Confederate brigadier
general

Inhabitants of the Ozark Mountains in
northwest Arkansas were almost solidly
against secession and the first peace society in the
Confederacy was organized there. The group of
dissenters had secret signs and passwords and
members pledged to encourage desertion from the
Confederate army and enlistment in the Union army.

Most of the Ozark hill people did not oppose slavery although few owned slaves. These mountain hill people simply did not see the question of slavery as sufficient cause to withdraw from the United States.

A February 1861 statewide election of delegates to a convention to consider secession produced 23,626 votes to remain in the Union and 17,927 votes to secede, and in mid-March the convention defeated an ordinance of secession on a 39 to 35 vote. But after the firing on Fort Sumter in mid-April, Lincoln called on the governor for troops to suppress the rebellion and the governor refused. The convention reconvened on May 6 and voted, with only one dissenting vote, to leave the Union.

The Peace and Constitutional Society was formed by men who opposed secession and refused to support the Confederacy. It had some 700 members in Van Buren, Newton, and Izard counties, and about 1,700 members in the whole state. Members were sworn to aid any member in any kind of distress, and to support the federal army when it came into the state. The penalty for disclosing any secrets of the order was death.

During 1862, the attempt to enforce the conscription law and the federal occupation of the northern part of the state increased opposition to the Confederacy. Some men hid out in the hills and others joined the federal army.

During the war, 8,789 white Arkansans enlisted in the Union army, including four infantry regiments, six cavalry regiments, and two artillery batteries, according to the state adjutant general. He added that the figure probably would exceed 10,000 if the Arkansans who fled the state and were enlisted in Missouri, Kansas, Iowa, Indiana, and Illinois regiments were counted. The Unionists also organized home guards for protection against Confederate forces.

One unusual project involved the establishment of fortified farming settlements in northwest Arkansas to allow famished refugees and other Unionists to attend to their spring planting. The commander of the First Arkansas Union Cavalry said that by April 1865, 16 settlements guarded and farmed by some 1,200 armed men had been established.

The First Arkansas U.S. Cavalry was formed in July 1862 and quickly moved to counter Confederate guerillas who were harassing Union sympathizers. However, the unit was caught in a rout at the Battle of Prairie Grove in December 1862, and the dispirited regiment was assigned to garrison duty in Fayetteville.

However, in April 1863, the regiment defended Fayetteville from a fierce Confederate attack and routed a force of 2,000 Confederates, greatly boosting morale. For the rest of the war, the First Arkansas skirmished with guerillas, conducted patrols, escorted wagon trains in northwest Arkansas, and protected the construction of a telegraph line from Fayetteville to Missouri.

Confederate home guards and regulars also were aggressive, rooting out and capturing nearly 200 Unionists, who were imprisoned in Little Rock and other locations. Some asked to take an oath of allegiance to the Confederacy and were enlisted in the rebel army. In November 1862, the state assembly passed a law making the giving of aid and comfort to the enemy a capital offense.

The Ozark uplands of non-slaveholding small farmers were fertile ground for organizing the "Mountain Feds," independent companies of Unionists operating under authority of the federal army. Among the most active was a unit formed by men seeking to avoid conscription and to protect themselves and their families from Confederate raids. Company B, First Arkansas Infantry Battalion, originated in Conway and Van Buren counties and was led by 51-year-old Thomas Jefferson (Jeff) Williams, a farmer and a Disciples of Christ preacher.

Elected captain, Williams, his four sons and extended family and neighbors formed the nucleus of the First Arkansas Infantry. Expecting to remain in their home territory in the Ozarks, the unit reluctantly moved with federal forces to the malarial swamps of eastern Arkansas where many died of disease and deprivation before the company was discharged in December 1862.

Following the capture of Little Rock in August 1863 and with the Arkansas River valley in Union hands by early September, the federals authorized the Union men of Conway County to form a company for the defense of their territory against Confederate guerillas. Again, Williams and his clan led the operation, returning to their home territory.

For nearly two years, Williams' Raiders, also called Williams Company of Scouts and Spies, scouted for the Third Arkansas Union Cavalry and fought a guerilla war against Confederate forces trying to regain control of north-central Arkansas.

But on Feb. 12, 1865, rebel guerillas surrounded the Williams cabin near present-day Center Ridge in Conway County and killed him with a volley of buckshot as he fled the cabin. Most of the rebel band was hunted down, killed or captured by Williams' Raiders. Williams' son Nathan took over command of the independent company.

Another band of anti-Confederates was based in Greasy Cove, a mountain pass at the head of the Little Missouri River. Composed of deserters and other dissidents, the band swept through the countryside, harassing Confederate loyalists and challenging Confederate authority.

The capture of Little Rock and the clearing of Confederate troops from the territory northeast of the Arkansas River spurred many Arkansans to hold meetings and publicly advocate a return to the Union.

Edward W. Gantt, a Hempstead County attorney, was elected to the U.S. House of Representatives in 1860, but did not take his seat as the secession crisis grew. A strong advocate of secession, he was commissioned a Confederate brigadier general and was captured with 7,000 other rebel soldiers in April 1862 on an island in the Mississippi near New Madrid, Missouri.

Released from prison in August 1862, Gantt became discouraged by his failure to secure another leadership post and by federal military advances in the western theater. He also became disillusioned and disgusted by what he saw as Jefferson Davis's incompetent conduct of the war, the failure of his military appointees, the imposition of martial law, and other attacks on civil liberties.

In June 1863, Gantt slipped to Vicksburg, Mississippi, and surrendered to Maj. Gen. Ulysses S. Grant. Meeting with President Lincoln in July 1863, he swore allegiance to the Union and promised his help in the war effort. In 1863 and 1864, he toured Northern states, speaking of pro-Union sentiment in the South and urging war-weary Northerners to persevere.

In his 1863 pamphlet in Favor of Reunion, Gantt aimed to assist the growth of the Unionist movement in Arkansas. He praised the state's Unionists for resisting the brutality of Confederate guerillas and military tribunals, and stressed the futility of further bloodletting and destruction for the sake of a lost cause.

"We thought we were fighting for constitutional liberty, when a tyrant was most mercilessly treading that constitution under his feet by every act of outrage that a conquered people can feel," Gantt wrote. "And, after nearly two years of strife, we awaken from a fearful baptism of blood to the terrible truth that the shadow of the despotism which we fled from under Mr. Lincoln dissolves into nothingness compared to the awful reign of tyranny that we have groaned under at the hands of Jefferson Davis and his minions."

"The sooner we lay down our arms and quit this hopeless struggle, the sooner our days of prosperity will return...Open the way for the return of husbands, fathers and sons, and bind up the broken links of the old Union," he implored.

Gantt worked with Lincoln to hammer out the details of the Ten Percent Plan under which Arkansas and other secessionist states could rejoin the Union when 10 percent of the eligible voters in 1860 swore allegiance to the Constitution and accented to emancipation. As the war ended in April 1865, he pushed for black political, legal, and economic rights as superintendent of the Arkansas Freedmen's Bureau.

Edward W. Gantt switched sides during the war. Courtesy of the Butler Center for Arkansas Studies.

After federal troops had occupied the state capital, Lincoln in March 1864 named Isaac Murphy provisional governor to start the state-making process. A school teacher, legislator, lawyer, and judge who served in the Union army until the occupation of Little Rock, Murphy's home was in Huntsville in northwestern Arkansas. While in Little Rock and absent from his home in Huntsville, Confederates raided and looted the town, giving special attention to the home of Murphy and the property of his family.

NORTH CAROLINA

Governor Vance called the law exempting affluent planters "perhaps the severest blow the Confederacy ever received" in terms of turning sentiment among non-slaveholders against the war effort.

T he main factor that kept North Carolina from quickly following other Southern states in secession was the large number of Unionists in the state, who soon became well-organized under the banner of the secret Order of the Heroes of America.

Unionist sentiment was strongest among non-slaveholding yeoman farmers in the western mountain region who felt they had little in common with the eastern slaveholding planters who largely dominated the state government. Unionists also were numerous along the coast and in the Piedmont "Quaker Belt."

While Quakers and others formed a minority that opposed slavery, Unionist politicians argued that remaining in the Union and avoiding civil war was the best route to preserving the South's "peculiar institution."

The election of Lincoln prompted secessionists to hold meetings across the state, and on Jan. 29, 1861, the General Assembly decided to put the question to a popular referendum on Feb. 28. The vote against secession was a close 47,323 to 46,672.

However, following the bombardment of Fort Sumter and Lincoln's call for 75,000 troops to "put down the rebellion," sentiment changed in North Carolina and the two other Upper South states, Virginia and Tennessee, which previously had rejected secession. The Virginia legislature's vote to secede on April 16 put added pressure on North Carolina.

Pro-secession Gov. John W. Ellis called a special secession of the General Assembly for May 1 and ordered seizure of all federal property. The Assembly voted May 13 to have an election for delegates to a secession convention to meet in Raleigh on May 20. The convention voted unanimously for secession and decided, at the request of Gov. Ellis, not to put the question to another popular vote.

The September 1862 election as governor of former U.S. Congressman Zebulon Vance, a staunch Unionist and reluctant secessionist by the time of the May 20 convention vote, was an indication of lingering Unionist sentiment in the state. Vance defeated a rabid secessionist by more than a 2-1 margin.

Nevertheless, the Tar Heel State became a bulwark of the Confederate cause, providing more men and supplies to the CSA and suffering more casualties than any other rebel state.

As was the case with the other Southern states, disaffection with the Confederacy grew as the war dragged on. The secret Order of the Heroes of America, which had some 10,000 members toward the end of the war, was strongest in the mountain west and in the Piedmont, whose Quaker and Moravian residents were pacifists and anti-slavery. Piedmont Unionists managed to elect peace candidates to the state legislature as well as a sheriff and other local officials.

The order was known as the "Red Strings" because of the red string worn in the coat label, an allusion to the Bible story of Rehab, who spied for the Hebrews.

Like the Peace Societies and other counterparts elsewhere in the South, the Heroes of America had secret passwords, handshakes, and oaths. They swore to protect deserters and escaped prisoners of war, to aid federal spy operations, and do whatever they could to undermine the Confederacy. The Heroes maintained close contact with federal forces, and established an "underground railroad" to enable Unionists to escape to federal lines.

Alexander H. Jones, a Heroes leader and editor of a Hendersonville newspaper, helped to inflame class resentments which the war had brought to the surface. "This great national strife originated with men and measures that were opposed to a democratic form of government. The fact is, these bombastic, high-falutin aristocratic fools have been in the habit of driving negroes and poor helpless white people until they think that they themselves are superior, (and) hate, deride and suspicion the poor."

Major anti-Confederate activity began on the eastern coast with the federal capture of forts and with hundreds of residents of Hatteras taking the oath of allegiance to the Union and promising to spy on rebel troop movements in exchange for federal army protection from Confederate vigilance committees. A Union military unit of some 60 men was organized and assigned to garrison duty. Secret Unionist meetings were held in Hyde, Washington, Tyrrell, and Beaufort counties bordering Pamlico Sound.

Maj. Gen. Ambrose Burnside captured Roanoke Island and New Bern in early 1862 and began to organize the First North Carolina Union Volunteers, who numbered 534 officers and men by January 1863, about half the size of a full regiment. This was followed by organization of the Second North Carolina Union Volunteers, which also failed to reach full regimental strength.

In December 1862, the First North Carolina Volunteers distinguished themselves in holding the town of Plymouth against an assault by the 17th North Carolina (CSA), and in November 1863 capturing a camp of rebels and taking 50 prisoners near Greenville.

In February 1864, Confederates under the command of Maj. Gen. George Pickett failed to capture New Bern, but took a number of First North Carolina prisoners—trying 22 of them as Confederate deserters and hanging them at Kinston. Demoralized by the executions and fearing the same fate, many of the regiment deserted during a successful April 1864 Confederate attack on Plymouth.

The demoralized eastern North Carolina recruits spent the rest of the war defending themselves against Confederate attacks and making occasional raids in conjunction with other Union military forces.

The Confederate conscription law of April 1862 created widespread disaffection and was seen as an admission that the government could not rely on voluntary enlistments. The law's substitution provision under which the well-to-do could pay others to enlist in their stead, followed by the "20 negro" exemption for planters enacted in October 1862, aroused bitter complaints of "a rich man's war and a poor man's fight."

Gov. Vance called that law exempting affluent planters "perhaps the severest blow the Confederacy ever received" in terms of turning sentiment among non-slaveholders against the war effort.

W.W. Holden, editor of the *Raleigh Standard*, and reputed to be a member of the Heroes of America, fanned the flames of discontent about these issues, barely avoiding arrest. The newspaper denounced Confederate President Jefferson Davis's imposition of martial law within a 10-mile radius of Confederate prisons, where both Union prisoners and suspected Unionists were held.

"The military go abroad, and without warrant or probable cause, they seize their prisoners, and with all possible speed, convey them to this place where the law is dead, and the sorrows of captivity fall on walls of stone," the newspaper complained. North Carolinians were urged to "denounce these attempts to betray our liberties and place us under a military despotism."

In September 1863, out-of-state Confederate soldiers passing through Raleigh ransacked the office of Holden's newspaper. In retaliation, two hundred of Holden's supporters ransacked the office of the pro-Jefferson Davis Raleigh *State Journal*. Moreover, Gov. Vance told Davis that the attack on the *Raleigh Standard* had aroused widespread indignation in the state and pleaded with Davis to prevent any similar incidents or risk seeing North Carolina troops in Lee's army rush back home "to the defense of their own State and her institutions."

By the spring of 1862, there were so many deserters in Chatham and Wilkes counties in the Piedmont and in Madison County in the mountain west that troops were sent to find and arrest them. However, many deserters, draft dodgers, and persons accused of disloyalty were released from the prison in Salisbury by state Chief Justice Richmond M. Pearson, who regarded the conscription law as unconstitutional under the writ of habeas corpus.

Pearson was supported by none other than Gov. Vance, who ordered militia officers "not to arrest persons who had been discharged under writ, and to resist such arrests by persons not authorized by a court having jurisdiction," thus encouraging desertion and draft evasion.

When the Confederate Congress passed the impressment and tax-in-kind laws on crops and livestock in the spring of 1863, falling most heavily on small farmers, Gov. Vance and editor Holden (the state's postwar Reconstruction governor), strongly complained.

However, Vance became so alarmed by Unionist activity and "the contagion of their example" in the mountain west—including robbing, burning, and plundering--that in January and April 1863 he asked the Confederate secretary of war to send troops. But anti-Confederate bands continued to grow in number and activity in western and central counties— forming a regiment, occupying a town, and executing conscription officers.

The Confederates conducted six military campaigns against the Unionists of Randolph County in the Piedmont from 1861 to 1865. To assist one campaign late in the war, Gen. Robert E. Lee sent about 500 of his seasoned troops to the county with orders to take no prisoners among those putting up armed resistance. Casualties were considerable on both sides of what amounted to guerilla warfare.

In January 1863, Confederate soldiers captured 15 men and boys, ages 14 to 60, who were accused of raiding salt and other supplies near Marshall in the western mountains. Although it was later reported that only five of the 15 were involved in the raid, 13 were executed by firing squad and buried in shallow graves. An historical marker of the "Shelton Laurel Massacre" was posted near Marshall.

Marker commemorating the Shelton Laurel Massacre in Marshall, NC. Public Domain.

Keith and Malinda Blalock were the only couple to fight side by side in both the Confederate and Union military. Keith, a staunch Unionist, allowed himself to be conscripted into the rebel army for the sake of Malinda's safety, planning to desert at the first opportunity. Malinda disguised herself as "Sam," Keith's brother, and served alongside her husband until a battle wound led to the discovery of her gender. Keith left with her for their Blue Ridge Mountains home, obtaining a medical discharge on false pretenses.

Later, after Confederate conscription officers tried to draft him, Keith and Malinda gathered up all the weapons they could carry and fled up the slopes of Grandfather Mountain. Others threatened with conscription soon followed them and formed a guerilla band that survived on the fish, game, and wild berries that were plentiful on the mountain.

After the Blalock band fired on a pursuing Home Guard company, wounding a soldier, Confederate authorities put a price on Keith Blalock's head. The couple fled to Tennessee where Blalock joined a Unionist guerilla outfit that served as scouts and raiders in the 10th Michigan Cavalry regiment with Maj. Gen. George Stoneman's rampaging "Cossacks" and which also piloted escaped federal soldiers and Confederate deserters to safety.

The legacy of the Blalocks is controversial to this day in the Blue Ridge counties of North Carolina where violence reigned a hundred and fifty years ago. Descendants of the couple's rebel enemies still revile them as traitors to the Confederate nation. And locals who count Unionists and federal soldiers among their ancestors regard them as loyal fighters for their country.

**Unionist guerilla Malinda Blalock.
Library of Congress.**

Another prominent Unionist partisan was
George Washington Kirk, who was raised in Greene
County, Tenn., but operated mainly in North
Carolina. Kirk enlisted in the Confederate army, but
deserted at the first opportunity. Along the
Tennessee-North Carolina border, he became a guide

for deserters and escaped Union prisoners of war who were seeking to reach federal lines.

Following Maj. Gen. Burnside's capture of Knoxville in September 1863, Kirk, now a 25-year-old Union army colonel, began enlisting western North Carolinians in Union regiments, the first being the Second North Carolina Mounted Volunteers. Later, Col. Kirk's Third North Carolina Mounted Volunteers raided Morganton, located beyond the Great Smoky Mountains, destroyed its military and railroad equipment, and captured a nearby Confederate camp for conscripts.

Kirk maintained constant pressure on Confederates in the western part of the state throughout the last two years of the war. He led forays into Caldwell, McDowell, Haywood, Watauga, and Macon counties before hostilities ended in May 1865. In February 1865, Kirk and his men pillaged the Smoky Mountains town of Waynesville , and on May 12 of that year he accepted the surrender of the last remnants of Thomas' Legion, a rebel unit made up mostly of Cherokees.

(In 1870, Kirk led federal troops to suppress a Ku Klux Klan insurrection in Alamance and Caswell counties in the north-central part of the state.)

In April 1863, Gen. Robert E. Lee reported "frequent desertions" from the North Carolina regiments. When Gen. Lee detailed a 50-man detachment from the 25th North Carolina regiment to return home and track down deserters, every man in the detachment deserted as well.

Gen. Dorsey Pender reported that his men were constantly receiving letters from home urging them to leave the army and promising that they would not be molested when they returned home, a promise reinforced by the Heroes of America, which vowed protection from the authorities. Most deserters headed for home, but some crossed into Union lines, providing intelligence and men to the federals.

Members of the Heroes of America included members of the state legislature, the superintendant of the North Carolina Railroad, the manager of the state salt works, and members of the army, seen with red strings of their lapels. Maj. Gen. William A. Smith, commander of the 37th battalion of the North Carolina Home Guard, was even reputed to be a member. Smith was elected by his men for promising that soldiers under his command would avoid direct conflict.

In the Battle of Gettysburg in early July 1863, North Carolina units of Lee's Army of Northern Virginia suffered huge losses. That July and August, about a hundred public meetings calling for peace talks with the Union took place in the state.

In the 1863 fall election, eight of the state's 10 elected congressmen had become opponents of the Davis administration and no original secessionist was elected to the Confederate Congress. Election results were similar in other rebel states.

At Flat Rock, near Asheville, three sisters of the Hollinger family, two of them married to rebel soldiers, regularly moved fugitives through Confederate lines to Union-held East Tennessee as part of an underground railroad.

In the increasingly safe haven of the Piedmont and mountain counties, North Carolina deserters were joined by those from South Carolina, Tennessee, and Georgia. In some communities, they managed to drive out all those in sympathy with the Confederacy. Union officers went among them to help organize scouting companies to provide intelligence, destroy enemy supply depots, and interrupt Confederate communications.

The Lumbee Indians of eastern North Carolina at first declared neutrality but became strongly pro-Union after the rebels began conscripting them for forced labor. Lumbee guerilla bands raided local plantations and supply depots and tore up rail lines.

Henry Berry Lowry led the most notable of these bands and came to be known as the Robin Hood of Robeson County. The Lowry Band served as guides for Maj. Gen. William T. Sherman's army when it entered North Carolina in March 1865.

In the western region, the Eastern Cherokee band from early 1862 formed Confederate battalions that ranged through the mountains enforcing conscription, impressing supplies, and rooting out Union sympathizers, which included some Cherokees.

But by 1863, food shortages and hunger stalked the Cherokee nation and hundreds of men deserted, some joining the federal army. One group of Cherokees aided Union Col. Kirk during his raid through the North Carolina mountains in June 1864, serving as Kirk's Third North Carolina Mounted Infantry Volunteers.

As food shortages grew, women's food riots broke out in towns across the Confederacy during the spring of 1863. In March, a crowd of some 50 hungry ax-wielding women in Salisbury stormed a government supply depot and stole 10 barrels of flour. Weeks later, food riots broke out in the Piedmont cities of Greensboro and High Point.

In the summer of 1864, about 20 soldiers' wives in Raleigh seized a government depot and most of its store of flour and grain. In the port city of Wilmington, women attacked a blockade runner and stole most of its cargo.

By the end of 1863, discontent and peace sentiment had become so strong that Gov. Vance told President Davis that an attempt should be made to negotiate with the Union. Davis replied that it could not be done because of the refusal of the Lincoln government.

In March 1864, Holden announced that he would be a candidate against Vance in the fall elections. Holden favored "peace now" negotiations between North Carolina and the Union while Vance advocated peace negotiations in cooperation with the other Confederate states. Elected representatives from North Carolina introduced peace resolutions in both the state legislature and the Confederate Congress.

A former ally of Vance, Holden was denounced as a traitor and a member of the Heroes of America. Vance won about 74 percent of the vote and Holden carried only the Piedmont counties of Randolph and Johnston. It was said that if ballot boxes had been placed in the woods of the western region, Holden would have received many more votes.

Vance worked to alleviate the increasing shortages and hardship among North Carolinians at the expense of the Confederate war effort. He appropriated the entire production of the state's cotton cloth, wool and shoe manufacturers for the state's soldiers and home population. He accumulated a large surplus of uniforms, shoes, and blankets for North Carolina's soldiers and refused to share it with Gen. Lee's ragged army.

By the early spring of 1865, as the end of the war drew close, secret meetings were being held in nearly all the counties west of the Blue Ridge for the purpose of choosing delegates to a convention that would organize a new state carved out of western North Carolina and eastern Tennessee.

By the end of the war, the provost marshal credited 3,156 North Carolinians in federal military service, not including uncounted numbers who enlisted in regiments of Tennessee, Kentucky, and Northern states. From 21 counties of western North Carolina alone, an estimated 4,000 men made their way to out-of-state Union regiments.

More than 5,000 North Carolina African Americans, mostly escaped slaves, joined the Union army during the war.

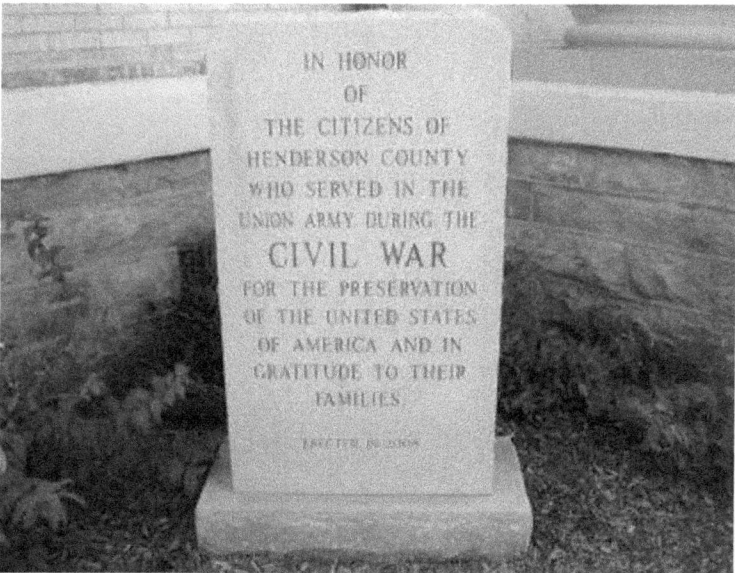

IN HONOR
OF
THE CITIZENS OF
HENDERSON COUNTY
WHO SERVED IN THE
UNION ARMY DURING THE
CIVIL WAR
FOR THE PRESERVATION
OF THE UNITED STATES
OF AMERICA AND IN
GRATITUDE TO THEIR
FAMILIES

Marker in western North Carolina Henderson County honors its Union soldiers.

TENNESSEE

Mostly from the eastern part of the state, more Tennessee men enlisted in federal military units during the war than did those from any other state of the Confederacy.

Tennessee was the last state to secede largely because of the East Tennessee mountaineer yeoman farmers who had little in common with the slaveholding planters to the west.

In the November 1860 presidential election, contested by four candidates, Tennesseans had voted for Constitutional Union Party candidate John Bell, a Tennessean and former U.S. Senator and U.S. House Speaker who had spoken out against secession for several years. (Virginia was the only other Southern state to vote for Bell, the rest voting for States Rights candidate John C. Breckenridge of Kentucky.)

As was the case with the two other states of the Upper South, the Battle of Fort Sumter and Lincoln's call for troops to put down the rebellion in mid-April 1861 sparked popular majorities to favor disunion. Prior to those events, a majority of Tennesseans opposed secession, voting 69,387 to 57,798 on Feb. 9, 1861, to remain in the Union.

But on May 9, the state General Assembly endorsed a "Declaration of Independence" favored by outspoken secessionist Gov. Isham Harris, to be followed by a popular referendum. Harris described the secession of the Southern states as a crisis caused by "long continued agitation on the slavery question" and "actual and threatened aggressions of the Northern states upon the well-defined constitutional rights of the Southern citizen."

Secession was strongly opposed by influential state leaders including U.S. Sen. Andrew Johnson, the only Southern senator who remained in his seat after secession, and William "Parson" Brownlow, a Methodist preacher and outspoken editor of the *Knoxville Whig*.

In response to the secession resolution,
Brownlow, Johnson and other leaders called a
Unionist convention in Knoxville for the last two
days of May, where 469 delegates from East
Tennessee rejected the resolution as "hasty,
inconsiderate, and unconstitutional," and "an act of
usurpation."

Brownlow declared, "We have no interest with
the Cotton States--we are a grain-growing and stock-
raising people. We can never live in a Southern
Confederacy and be made the hewers of wood and
drawers of water for a set of aristocrats and
overbearing tyrants."

Fiery Unionist "Parson" William Brownlow became governor of Tennessee. Public domain.

In the popular referendum June 8, the vote was more than 2 to 1 for secession—102,172 to 47,238. But in East Tennessee, the vote was well over 2 to 1 against secession—32,923 to 14,780. East Tennessee leaders charged the Nashville-based secessionists with suppression of speech, fraud, and voter intimidation in the state's western region.

Nine days after the referendum, East Tennessee Unionists from 26 counties held a convention in Johnson's hometown of Greenevile in the far northeastern tip of the state. Meanwhile, Johnson, who had campaigned vigorously against secession in the referendum, had fled north to Ohio following assassination threats. The convention voted to petition Nashville for the right to separate statehood for East Tennessee. But the General Assembly, predictably, rejected the petition, and Gov. Harris dispatched troops to occupy the region and prevent secession from the state.

Late in 1861, the county court of Scott County, whose citizens had voted against secession by a 521-19 majority (the highest percentage of any county), passed a resolution that it was separating from Tennessee and forming the "Free and Independent State of Scott."

Monument to Union soldiers from Greene County. Courtesy of Brent Moore.

Local Unionist guerilla bands began organizing for a coordinated attack in December on nine bridges of the East Tennessee rail network, expecting a simultaneous Union invasion that failed to materialize. Five of the nine bridges were destroyed, but the absence of the promised Union forces aborted a planned Unionist uprising and led to the arrest and imprisonment of some 400 Union sympathizers.

Several of the prisoners were hanged and two were left hanging for two days in Greeneville, on an oak tree close to the railroad track where the trains passed by slowly so that passengers could kick and strike their dead bodies with canes.

Among those jailed for the bridge-burning was the Rev. Brownlow, who was not directly involved in the incident. Threatened with execution on a charge of treason, a brigadier general in charge of the prison offered to release Brownlow if he would swear an oath of allegiance to the Confederacy.

The parson replied, "I intend to lie here until I rot from disease, or die of old age before I would take such an oath. I deny your right to administer such an oath. I deny that you have any government other than a big Southern mob. You have never been recognized by any civilized Power on the face of the earth, and you never will be. "

Instead of being hanged, Brownlow was banished to the North after being escorted to Union lines under a flag of truce. He embarked on a speaking tour of Northern cities and wrote a book—
Sketches of the Rise, Progress, and Decline of Secession;

with a Narrative of Personal Adventures Among the Rebels, which immediately became a best seller in the North, and clandestine copies were circulated in the South.

He wrote that Southern Unionists "have no respect or confidence in any Northern man who sympathizes with this infernal rebellion–nor should any be tolerated." He portrayed the Confederacy as a "murderous and bloodthirsty" despotism ruled by scheming demagogues and policed by thugs and drunks.

"Hanging is going on all over East Tennessee," he wrote. "They shoot them down in the fields—they whip them, in the counties of Campbell and Anderson they actually lacerate with switches the bodies of females, wives and daughters of Union men. They show no quarter to male or female. They rob their houses and they throw them into prison. Our jails are full now, and we have complained and thought it hard that our Government has not come to our relief, for a more loyal, a more devoted people to the Stars and Stripes never lived on the face of God's earth than the Union people of Tennessee."

In September 1863, Brownlow returned to Knoxville in the company of Union Maj. Gen.

Ambrose Burnside and started up his old newspaper. Burnside's arrival was greeted by a joyful crowd of Union supporters.

Brownlow's son John, meanwhile, was commissioned colonel of the 9th Tennessee U.S. Cavalry.

Kirby Smith, commanding the Confederate department of East Tennessee, complained in March 1862 that "every effort made by the state authorities to call out the militia of East Tennessee has proved unavailing."

An East Tennessee militia captain, ordered to report with half of his command to Confederate headquarters at Monroe, instead fled to Union lines in Kentucky with 44 of his men and promised that many more were on their way.

Confederate home guards often showed no mercy hunting down deserters and suspected Unionists in Tennessee and throughout the South. Home guards that patrolled the far northeast counties of Carter, Johnson, and Sullivan were especially brutal. It was said they would rather shoot Union men than squirrels—they "loved to see them jump." Their killings on the Carter-Sullivan line were so numerous that locals called it "the boneyard."

Mostly from the eastern part of the state, more Tennessee men enlisted in federal military units during the war than did those from any other state of the Confederacy.

According to the state adjutant general, 22,960 whites and 5,201 blacks were mustered into Tennessee regiments of the U.S. military. Beginning in September 1864, 15,400 whites and 3,834 blacks were enrolled in local militia and home guard units. In the course of the war, some of these units were mustered into the U.S. military. In addition, an estimated 7,000 men, most of them fleeing to Union lines in Kentucky, enlisted in regiments from other states.

As of Jan. 1, 1865, the adjutant general listed 32 white regiments, including the 1st through the 14th Union Cavalry, the 1st through the 11th Infantry, the 1st through 7th Mounted Infantry, and five separately numbered "colored" regiments.

Major General Ambrose Burnside's triumphal arrival in Knoxville. Public domain.

The 1st and 2nd East Tennessee Infantry during the winter of 1861-62 tried to drive the rebels from East Tennessee as part of Brig. Gen. George W. Morgan's division of the Army of Ohio, but with little success.

During the Stones River campaign from Dec. 26, 1862 to Jan. 5, 1863, in the bloody Second Battle of Murfreesboro, the First Tennessee, part of Maj. Gen. William Rosecrans' Army of the Cumberland, greatly distinguished itself, twice charging and driving from the field the Confederates under the command of Maj. Gen. Braxton Bragg's Army of Tennessee.

During Confederate Lt. Gen. Nathan Bedford Forrest's foray into West Tennessee and Kentucky during the spring of 1864, the 13th West Tennessee U.S. Cavalry was decimated at the Battle of Fort Pillow on the Mississippi, where much of the garrison of 557 men was massacred after the fort was captured. The rebels singled out for slaughter the fort's black soldiers and its "home-made Yankees" (native white Unionists). After Forrest's raid, however, the 10th and 12th Tennessee U.S. Cavalry pursued and harassed his cavalry for over 200 miles.

When the Confederates under Maj. Gen. John Bell Hood attempted to retake Nashville in the fall of 1864, they were repulsed by federal forces led by Maj. Gen. George H. Thomas and including four Tennessee infantry regiments, six cavalry regiments, and two batteries of light artillery.

During the climactic mid-December Battle of Nashville, the 1st Tennessee Cavalry captured a battle flag of one of Forrest's divisions while the 12th Cavalry captured three of the division's battle flags along with a brigadier general, broke its line of battle, and scattered its men in all directions.

During the war's final months, Union Tennessee troops finished the job of subduing the rebels in East Tennessee and western North Carolina. Involved in this task were four cavalry regiments, four infantry regiments, and a light artillery regiment.

Unionist fighters were not confined to East Tennessee. Fielding Hurst, the largest landowner in McNairy County, was arrested in 1861 for his outspoken Unionist sentiments and imprisoned in Nashville along with other Union sympathizers. Freed by the federal occupation of Nashville, he returned to McNairy County in southwest-central Tennessee where he organized a unit of mounted scouts in what became known as the "Hurst Nation."

Hurst's unit patrolled the countryside, skirmished with Confederates, and provided intelligence to the federal army. Given a commission by Andrew Johnson, the state's military governor, his unit was designated the 6th Tennessee Cavalry (USA), which successfully pursued Confederate guerillas and was pursued by Forrest's cavalry. An historical marker in Bethel Springs commemorates Hurst.

Marker commemorating the "Hurst Nation" an extended family of Unionist guerilla cavalry, in northwest McNairy County. Public domain.

Robert W. Boone, a great-grandson of Daniel Boone, was in the secret service of the Union army, serving undercover in Confederate regiments, operating as a scout and spy, and piloting escaping federal prisoners to Union lines.

Tennessee women also aided the Union cause, most prominent being Jeannette Mabry of Knox County, who was married to a Confederate colonel. During the guerilla war that prevailed in the Smoky Mountains, Mabry was an inspiration to other loyalists. At great risk to herself, she was in constant contact with Union guides and envoys from the federal lines, and she organized assistance to indigent refugees.

Another Unionist, Lucy Williams, was responsible for the death of Confederate raider and general John Hunt Morgan in early September 1864. Williams was residing at the home of her prominent Confederate mother-in-law in Greeneville when she learned that Morgan and his staff would be spending the night there. She learned of a planned attack by Morgan and slipped away, riding her horse 15 miles through heavy rain to Union lines where she shared her information. Early the next morning, federal troops surrounded the Williams home and killed Morgan as he tried to escape.

And Unionist Pauline Cushman, a New Orleans-born actress, established her secessionist credentials by toasting Jefferson Davis onstage in Union-occupied Louisville, Kentucky. In Nashville, Cushman gained the confidence of prominent secessionists while working with the Union Army of the Cumberland, sharing military intelligence. She also acted as a courier and her risky exploits ranged from Tennessee and Kentucky to the mountainous northern tiers of Mississippi, Alabama, and Georgia.

Pauline Cushman, actress and spy. Public domain.

Cushman was caught in the spring of 1863 near Shelbyville, Tennessee, with drawings she had taken from a Confederate army engineer. She escaped, was recaptured, and was sentenced to death. Three days before her scheduled execution, she was rescued by

Union forces. General James A. Garfield (later president) gave her an honorary commission as a major, and she spent the rest of the war touring the North and describing her exploits as the Little Major.

Many Confederate soldiers from East Tennessee were unreliable, even forming a Peace Society, and joining the Order of the Heroes of America, a secret Unionist society that was based in North Carolina. These soldiers were seldom assigned to picket duty for fear of desertion. The Confederate defeats at Gettysburg and Vicksburg in early July 1863 greatly increased the ranks of deserters, many of whom enlisted in the Union army.

Although the lowland counties of West Tennessee generally favored secession, several counties on the western side of the Tennessee River contained considerable numbers of Unionists. In the June 1861 referendum, a majority of voters in five eastern West Tennessee counties—Weakley, Carroll, Henderson, Decatur, and Hardin—voted against secession despite intimidation at the polls. These and other counties in the region contributed three new regiments and parts of several other regiments to the federal army.

When Gen. Ulysses S. Grant in February 1862 captured Forts Henry and Donelson in the proximity of eastern West Tennessee, Unionists took the opportunity to volunteer in new federal regiments— the 6th and 7th Tennessee Cavalries, and the 2nd Tennessee Mounted Infantry. The regiments guarded railroads, logistical routes, and lines of communication, and also served as guides and spies, fought rebel guerillas, and commandeered provisions for the Union army.

Confederate general Forrest captured Union City in the northwest tip of the state in March 1864, taking prisoner much of the 7th Tennessee, and sending them to the notorious Andersonville prison in Georgia where two-thirds of them died of disease and starvation.

As Grant's army steamed up the Tennessee River after his capture of the two key forts, crowds of citizens cheered, clapped, and waved American flags as his troops passed by. More than 4,000 slaves in the area sought protection behind federal lines, providing manual labor and useful intelligence to the federals. Slave desertions from plantations in Tennessee and throughout the South helped the Union cause and deprived the Confederacy of sorely needed economic manpower.

When Burnside entered Knoxville in September 1863, Unionists gave him an enthusiastic welcome. But when Confederate general James Longstreet shut up Burnside inside the city, cutting off food and supplies, his army was faced with starvation. Unionists came to the rescue by floating at night small boats filled with food down the Tennessee River into the city.

Democrat Johnson, following his service in the Senate from 1857 to 1862, was appointed military governor of the occupied portion of Tennessee. In this position, his speeches and pamphlets, which received wide distribution and discussion in the North and Upper South, attacked the Southern slaveholding aristocracy and argued that under the U.S. Constitution, "a state cannot commit suicide" by seceding from the Union.

Johnson declared in a January1864 speech that "traitors must be punished and impoverished. Their social power must be destroyed, and the effects that give them power and influence must be taken away."

He said Tennessee's return to the Union depended on its support for the abolition of slavery, whose beneficiaries would be the state's small, independent farmers who ought to be given land by cutting up large cotton plantations into small plots. Earlier in the war, Lincoln's Jan. 1, 1863 Emancipation Proclamation had exempted Tennessee for the purpose of not alienating pro-slavery Unionists.

Johnson said the end of slavery "is not extended to the negro only, for this will free more white men than it will black men." He added, however, "I am for a white man's government, and in favor of free white qualified voters controlling this county, without regard to negroes." He expressed the hope that ex-slaves would be transplanted to Mexico or Africa, or "some other country congenial to his nature, where there is not that difference in class or distinction, in reference to blood or color."

Johnson urged the election of citizens, after swearing allegiance to the Union, first to local offices in the federally controlled parts of the state, and then to a state convention to seek readmission to the Union.

A Unionist state government was established in 1865 and "Parson" Brownlow was elected governor. Originally a staunch defender of slavery, Brownlow embraced the 13th Amendment that abolished it and even allied himself with the Radical Republicans during Reconstruction. Brownlow helped Tennessee to become the first ex-Confederate state to be readmitted to the Union on July 24, 1866.

POSTSCRIPT

The Civil War lingers on in radically divergent memories of its causes, the reasons for the Confederacy's defeat, the nature of Reconstruction, and the legacy left by our greatest national trauma. That trauma was especially acute in the defeated and war-ravaged South and has affected Southerners' views of the Lost Cause to this day.

There certainly is room for debate regarding the extent to which opposition to the Confederacy among Southerners affected the course of the war. The collapse of the German home front during World War I and its decisive effect on ending the war has been well documented. By comparison, the erosion of Southern support for the war was far more subtle and more gradual in the role it played in ending the fighting.

For those who wish to dig deeper into the events and people described in the book, the works listed in bibliography should be explored.

Southerners whose ancestors fought for the Confederacy may rightfully take pride in their courage and sacrifice. Southerners who ancestors fought for the Union as soldiers, guerillas or in underground networks also may take pride in their role in preserving the United States and ending slavery. As noted in the book, memorials to these Unionists have been erected in several Southern states, some of them marking the war's 150th anniversary.

In the light of the facts presented in this book and other works listed in the bibliography, those who cling to the view of a solid Confederate South face a tough uphill battle to convince others. But the question is still open for debate and those wishing to participate in the discussion are invited to do so at http://www.facebook.com/DividedWeFallBook

ABOUT THE AUTHOR

Calvin Goddard Zon's first book—The Good
Fight That Didn't End: Henry P. Goddard's Accounts
of Civil War and Peace—was published by the
University of South Carolina Press in 2008. The
critically acclaimed book is based on the writings of
his great-grandfather, a captain in the 14th
Connecticut Infantry. Zon is a Washington, D.C.-
based historian and journalist, having been a staff
writer for the Washington Star daily, Press

Associates, Inc., and the United Mine Workers of America Journal. He has written for the Civil War News, the Civil War Times, People, The Progressive, National Catholic Reporter, and In These Times. He earned a B.A. from Davidson College, where he majored in American history, and received an M.A. from the American University. He has taught high school American history. He is past commander of Lincoln-Cushing Camp No. 2 of the Sons of Union Veterans of the Civil War, and a companion of the Military Order of the Loyal Legion of the United States. He served for six years in the U.S. Army Reserve. He is a member of the Lincoln Forum and a board member of the Lincoln at the Crossroads Alliance. He has presented his first book at Civil War Roundtables from Connecticut to South Carolina. One of his Virginia ancestors served on the staff of Stonewall Jackson.

BIBLIOGRAPHY

Ash, Steven V., editor, *Secessionists and Other Scoundrels: Selections from Parson Brownlow's Book*, Louisiana State University Press, 1999.

Barnes, Kenneth C., *Encyclopedia of Arkansas History & Culture*, 2013.

Barrett, John G. *The Civil War in North Carolina*, 1963.

Baum, Dale, *The Shattering of Texas Unionism*, Louisiana State University Press, 1998.

Current, Richard Nelson, *Lincoln's Loyalists: Unions Soldiers from the Confederacy*, Northeastern University Press, 1992.

Dyer, Thomas G., Secret *Yankee: The Union Circle in Confederate Atlanta*, Johns Hopkins University Press, 1999.

Encyclopedia of Arkansas History and Culture, "Jeff Williams," 2014.

Escott, Paul D., After *Secession: Jefferson Davis and the Failure of Confederate Nationalism*, Louisiana State University Press, 1978.

Faulkner, Ronnie W., "Secession," North Carolina History Project.

Foner, Eric, "The South's Inner Civil War," *Civil War and Reconstruction*, 1988.

Freehling, Wiliam W., *The South vs. The South: How Anti-Confederate Southerners Shaped the Course of the Civil War*, 2001.

Frisby, Derek W., "Fielding Hurst," Tennessee Encyclopedia of History and Culture.

History Department of North Carolina State University. "Why Was North Carolina Reluctant to Secede from the Union?" 2014.

Holley, Peggy Scott, "Unionists in Eastern West Tennessee, 1861-1865," Austin, TX, 2004.

Hurst, Robert, *The Southern Loyalist*, Swannco.net, 2014.

Inscoe, John C. and Kenzer, Robert C., *Enemies of the Country: New Perspectives on Unionists in the Civil War South*, University of Georgia Press, 2001.

Jones, Wilmer L., *Behind Enemy Lines: Civil War Spies, Raiders, and Guerillas*, Taylor Publishing Company, 20001.

Levine, Bruce, *The Fall of the House of Dixie*, Random House, 2013.

Louisiana State Museum *Cabildo and Louisiana History*, 2014.

Murphree, R. Boyd, "Florida and the Civil War: A Short History," *A Guide to Civil War Records at the State Archives of Florida*, 2014.

Museum of Florida History, "Florida in the Civil War," 2014.

Neely Jr., Mark E., Southern *Rights: Political Prisoners and the Myth of Confederate Constitutionalism*, University Press of Virginia, 1999.

North Carolina Museum of History, "North Carolina and the Civil War, The Home Front," 2005.

Parker, Richard, and Boyd, Emily, "The Great Hanging at Gainesville," *The New York Times*, Oct. 16, 2012.

Pratt, Adam, "Unionism in Louisiana," *In KnowLA Encyclopedia of Louisiana,* edited by David Johnson, Louisiana Endowment for the Humanities, 2013.

Stevens, Peter F., *Rebels in Blue: The Story of Keith and Malinda Blalock,* Taylor Publishing Company, 2000.

Tatum, Georgia Lee, *Disloyalty in the Confederacy,* University of Nebraska Press, 2000.

Texas State Historical Association, "The Great Hanging at Gainesville," *Civil War History* 22, 1976.

Trelease, Allen W., "G.W. Kirk," 1988, NCpedia.

Tinkler, Robert, Inside Chico State "Southern Unionists in the Civil War," 2012.

Underwood, Rodman L., *Death on the Nueces: German Texans Treue der Union,* Eakin Press, 2000.

Varon, Elizabeth R., *The True Story of Elizabeth Van Lew, A Union Agent in the Heart of the Confederacy,* Oxford University Press, 2003.

Wakelyn, Jon L., *Southern Unionist Pamphlets and the Civil War,* University of Missouri Press, 1999.

Williams, David, *Bitterly Divided: The South's Inner Civil War,* The New Press, 2008.

Wilson, William Moss, "East Tennessee's Unruly Unionists," *The New York Times*, June 16, 2011.

Wolfe, Brendan, "Unionism in Virginia During the Civil War," *Encyclopedia Virginia*, 20